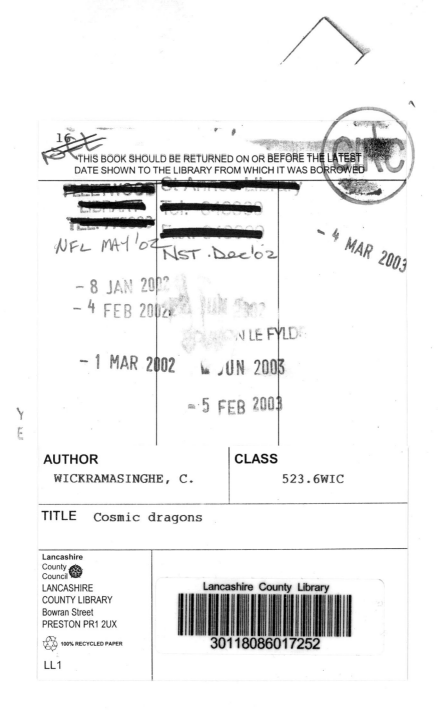

AUTHOR	CLASS
WICKRAMASINGHE, C.	523.6WIC

TITLE   Cosmic dragons

COSMIC DRAGONS

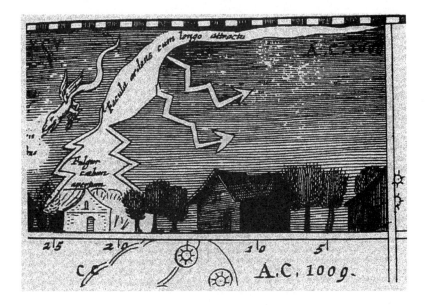

Frontispiece   Dragon motif depicting a cometary missile attack,
from an early manuscript

COSMIC DRAGONS
Life and Death on our Planet

Chandra Wickramasinghe

Souvenir Press

This Edition published 2001 by
Souvenir Press Ltd.,
43 Great Russell Street, London WC1B 3PD

ISBN 0 285 63606 5

Printed in Great Britain by
Creative print and Design Group (Wales), Ebbw Vale

For Janaki, Kamala and Anil –
May you see a better world,
in which Truth, Tolerance and Compassion prevail . . .

Whatever lamps on Earth or Heaven may shine
Are portions of one power which is mine.

P.B. Shelley, *Hymn to Apollo*

Contents

Prologue

This book is about cosmic dragons. The mythological dragon, just like comets of old, was universally feared as well as venerated throughout the ancient world. The oldest traditions of China had the dragon (*lung*) primarily as a beneficent power dispensing blessings. Dragons were supposed to move through the heavens gathering clouds and moisture and dispensing life-giving rain. But like rain they can on occasion be destructive. In China this negative power was never emphasised. It was accepted as being part of their *ying yang* character, contributing to an overwhelming might, and it is perhaps for this reason that the dragon was adopted as the symbol of Chinese imperial power. In Europe, however, it was the malevolent character of the dragon that came to be emphasised. The dragon became an alien, hostile, pagan force of evil to be checked, and so it came about that the dragon had to be slain by heroes such as Saint George. In this book we adopt the ancient Chinese dragon to represent the comet. As we shall see, comets dispense life throughout the cosmos, but on occasion they can also kill.

In astronomy Draco, the dragon is an ancient constellation in the northern skies: a group of six bright stars bearing the outline of a serpent. The star Alpha Draconis is notable as being the Pole Star at the time when the Great Pyramid of Giza was built at around 2500BC; and even more remarkable is the fact that one of the tunnels of this pyramid points precisely to where the star would have been in the ancient sky (see Chapter 8). Could it have been that cosmic dragons of a malevolent kind appeared to radiate from a point in the constellation of Draco around 2500BC?

The idea that living microorganisms dispensed by our cosmic dragons might be floating in the unfathomed depths of space may sound bizarre and incredible, but if true is indisputably fantastic. Likewise, the thought of cosmic dragons darting through our skies, whipping their tails where they will, causing disaster to life on planets may sound a travesty of common sense. But if such depictions represent even an approximation to the truth, the imagery of cosmic dragons will not have been overdone. In this book we show how metaphor has turned to fact.

1. Astronomical Backdrop

Awake! for morning in Bowl of Night
Has flung the stone that puts the Stars to Flight
And lo! the Hunter of the East has caught
The Sultan's Turret in a Noose of Light.

Edward FitzGerald,
Rubaiyat of Omar Khayyam (1809–1883)

The year AD2001 came and went, quietly, almost without incident, like water under a bridge. A thousand years earlier the dawn of a new millennium would have been riddled with fears of Armageddon, prophecies of the end of the world. Such fears were obviously not realised, but nor have they entirely disappeared today. Every now and then one hears rumblings of an asteroid or comet heading towards our planet on a possible collision course. So far such scares have turned out to be premature false alarms by impetuous scientists using incomplete evidence. But the risk of collisions still looms large, albeit on a distant horizon, and there is now a more or less general acceptance of this fact. It is only a matter of time before human society will have to brace itself for such an inevitable contingency. When the alarm bells finally ring out decisions will have to be made as to what course of action to pursue: to destroy the incoming culprit missile before it strikes; to deflect it somehow; or to accept the inevitability of a collision, and then to alleviate its worst effect. Along with earthquakes and volcanoes, impacts

1

which, as we shall see later, have punctuated history, might come to be regarded as yet another natural hazard that has to be faced.

In our planet's distant past collisions with comets were not always bad. One might say with some justification that they have been a mixed blessing. We shall see (in Chapter 7) that such collisions had the effect of moulding our primitive Earth into a bio-friendly planet. They may have been responsible for bringing life on to the planet in the first place, but intermittently at later times they also caused the extinction of species.

One of the main concerns in this book will be to trace our cosmic heritage. To answer such questions as 'Where did we come from?' 'Is there life elsewhere in the Universe?' 'When and where did life on Earth begin?' 'What will be the eventual fate of life on our planet?' These are age-old questions that have occupied the minds of our ancestors from time immemorial. They are also amongst the most important questions of modern science.

It is only now that serious attempts are being made to answer these questions, with all the best resources of modern space technologies being channelled to this end.

Now that the threat of a future episode of comet and asteroid impacts is being taken seriously, plans are afoot to deploy space technology to probe the nature of this threat more carefully. A database of potentially hazardous objects in the near-Earth environment is being diligently assembled, and asteroids and comets are being investigated to assess the nature of the threat. The NASA space probe called NEAR was launched in 1996 to the 34-kilometre long sausage-shaped asteroid Eros, and in February 2001 this was made to crash on to its pock-marked surface (see Chapter 9). The information gleaned in the seconds prior to the crash would yield vital data as to how strong or fragile asteroidal material really is, thus helping to plan for dealing with an impact by a similar object.

Apart from NEAR a fleet of other space probes and spacecraft are poised to explore the planets and satellites of our solar system, including comets, with ever increasing levels of sophistication. In February 1999 one such space probe called Stardust set out on a seven-year voyage to a comet called Wildt 2. Its mission is far-reaching – to go to a comet and bring back to us what it contained. The eventual return of material scheduled for 2006 is still a few years away, but already studies of cosmic dust that Stardust has encountered *en route* to the comet has revealed some tantalising evidence of life beyond the Earth.

What Stardust did was essentially very simple. Travelling far outside the plane of the planets, at 65,000 miles per hour, it put out an arm with a detector at the end of it. Five dust particles from interstellar space smashed on to the detector at a speed of 30 kilometres per second. Then Stardust looked at the shattered fragments and analysed what they are. The fragments turned out to be extremely large, highly complex organic molecules arranged in three-dimensional frameworks of cross-linked structures. These are uncannily similar to structures that make up the cell walls of bacteria – the strongest components of bacteria that could survive such violent impacts.

Well, what does this all mean? Has Stardust discovered bacteria and life in interstellar space? May be. Although from this one single piece of evidence one cannot be absolutely certain of course. What is certain is that the search is surely on, a search that we shall see had already started many years ago. The story of the search is the subject of this book.

Let's begin at the beginning by exploring the history of the matter from which the cosmic dragons – the comets – and all living and nonliving things are made. Take a rock and hammer it with sufficient force, and you will smash it into smithereens. Hammer these pieces hard enough and you will break them down into even smaller pieces. You can contemplate successive division and sub-division in this way, provided that at each stage an appropriate kind of 'hammer' is used. But how far can this process be continued? How small will the ultimate units of matter turn out to be?

These questions were already being asked by the ancient Greeks in the fifth century BC. The Greek philosopher Democritus (*ca.* 460–370BC) postulated the existence of atoms as the smallest units of a substance. The Greeks invented the concept of atoms because they found it impossible to contemplate an indefinite divisibility of matter. The details of their atomic theories were of course crude by our modern standards – for example, a sugar atom was said to be white and sweet. But the underlying philosophy of these ancient concepts was beyond reproach. Lucretius (96–55BC) modernised the older ideas considerably, He supposed for instance that water 'atoms' disappear from a surface in the process of evaporation because they fly off like tiny bullets, anticipating the modern theory of gas kinetics.

Today physicists are still grappling with the same age-old basic questions that were posed by the Greeks 2,500 years ago. The quest continues for the smallest units into which matter might be ultimately divisible. In the nineteenth century it was thought that an ultimate unit was discovered and this was called the atom. Not 'atoms' of sugar or water, but true atoms such as hydrogen, oxygen, carbon and so on. Then in the early part of the twentieth century the atom was split. It was subdivided into electrons, protons and neutrons, and nowadays there are even smaller units known as quarks.

In this book we shall for the most part not proceed with this subdivision process beyond electrons, protons and neutrons, and of these, the latter two – protons and neutrons – will be the most relevant for our purposes here. Assemblages of protons and neutrons make up atomic nuclei, and it is the atomic nuclei themselves that are crucial for the definition of the everyday materials that we deal with. Atoms of any particular kind are comprised of tightly packed nuclei made of protons and neutrons surrounded by clouds of orbiting electrons. It is the electrons that take part in chemistry and chemical reactions.

There are more than 90 different varieties of atoms and atomic nuclei in the world around us: hydrogen, helium, carbon, nitrogen, oxygen, phosphorous, calcium, magnesium and iron are a few examples. Our experience of these atoms is mainly derived through the molecular structures into which they get assembled: hydrogen (H) and oxygen (O) combine to form water (H_2O); sodium (Na) and chlorine (Cl) combine to form a molecule of common salt (NaCl). The green colouring substance of plants, chlorophyll, is a molecule made up of over 100 individual atoms, while the molecules that make up our genes are comprised tens of thousands of atoms. We shall show that the genesis of the atoms that make up molecules, that make up genes, that make up the entire spectrum of life is writ in the stars. Our genetic ancestry at the deepest level is buried in the vast cosmos that surrounds us.

In days gone by, before there were connurbations and city lights, our ancestors would have been able to enjoy a deep and meaningful relationship with the universe. The power of the cosmos would have asserted itself with unmistakable clarity. Nowadays our perceptions of our cosmic ancestry are blurred not only by light pollution but also by centuries of prejudice. When we look up at the night sky we are lumbered with time-hallowed intellectual constructs that on the whole tend to devalue the connection that undoubtedly exists between life on Earth and the wider universe. This book seeks to rediscover the lost connection and examine its importance in relation to our understanding of the Universe as whole.

The starting point of our story must be the spectacle of the night sky itself, as much of it as it is possible to retrieve in modern times. In the few remaining unspoilt spots on the planet – on a cruise-ship on the high seas, in the remotest depths of Jamaica or Sri Lanka, for example – the ancient splendour of the heavens can still be enjoyed. On a moonless night in the darkest available location, a swathe of whitish light can be seen to stretch across the sky from horizon to horizon, interspersed with bright stars and the occasional complex of dark clouds. Such a sight would have intrigued our ancestors for millennia and given rise to numerous myths and legends. The vault of the night sky appearing like a stream of diffuse white light led the ancient Greeks to describe it as a river of milk flowing forth from the breast of Hera, wife of Zeus, the King of the gods. The word galaxy is in fact derived from the Greek word for milk. The Romans too saw this vault of the heavens as a milky path across the sky – the *via Lactea*, or Milky Way.

The idea of a luminous 'celestial fluid' straddling the heavens persisted right through to the time of Galileo (1564–1642). In 1610 Galileo Galilei turned his telescope on the Milky Way and showed it to comprise vast numbers of faint stars that could not be resolved with the naked eye.

This magnificent celestial spectacle is by no means static. On a moonless night every hour a few shooting stars could be seen flitting across the sky, each one starting mysteriously from a point in the sky and ending its fleeting life as an incandescent streak. These are of course the meteors, and their apparent great distances and presence amidst the stars is an illusion. Every single meteor is a piece of debris from our own planetary system with a size in the range of millimetres to centimetres, plunging into the Earth's atmosphere and burning up at heights of tens of kilometres above the surface. There would also be much smaller particles in the train of meteoric débris, particles measuring hundredths of a millimetre. The smallest of such particles reach the ground without being burned. In November 1999 an incoming meteor train (the Leonids) was found to contain signatures of organic material that was described euphemistically as 'building blocks of life'. Also the trail of these meteors was found to be luminescent for several seconds, similar to the behaviour of bioluminescent bacteria as they are burnt.

The meteors we see on an average night, a few per hour, turn into showers of several to many events per minute on certain days of the year when the Earth in its yearly path round the Sun encounters tubes of concentrated debris. Such tubes arise because comets shed particulate material, and such material, like the comets themselves, travel in orbits around the Sun. Each comet leaves behind its own orbital tube of debris that the Earth crosses at particular dates in the year. For instance Comet Encke gives rise to the Taurid meteor stream that is crossed every November and June, and Comet Giacobini-Zinner gives rise to the Draconids that are crossed in early October. Rarely, the material in a meteor stream is so dense that for a few hours the hourly meteor rate can shoot up to very high values. Such meteor storms have been recorded for the Draconids in 1933, with an hourly rate of 20,000, and for the Leonids in 1966 with an hourly rate of 150,000.

The average amount of particulate matter from comets, now known to be organic, that enters the Earth's atmosphere on a daily basis, is measured in hundreds of tonnes. The thesis to be developed in later chapters is that this input of material from comets carries the potential for all life, and the capacity for destroying it as well.

In oriental folklore shooting stars or meteors together with fire-flies on the ground are often regarded as messengers from the stars. The missionary link may not be entirely fictional after all. Whenever you watch the shimmering spectacle of the night sky, with meteors flitting against the background of billions of stars like the sun in the Milky Way, have you ever thought for a moment how many other living creatures there might be watching a very similar scene? How many planetary systems like our own solar system might exist throughout the cosmos, and how many billions of Earth-like planets replete with the intelligent life might there be orbiting distant stars? How do stars, planets and life come into being?

In modern times when exotic areas of cosmic physics – black holes, worm-holes and the like – clamour for public attention and the public purse, the old-fashioned life story of stars and planets tends to be relegated to the back burner. Yet, as we shall see, this is precisely where our cosmic heritage and destiny would seem to lie. Dust to dust . . . Born of stardust, to stardust to return.

Amongst the seething mass of stars in the Milky Way – *via Lactea* – our ancestors saw patterns of life. The brightest visible stars were taken to define constellations of which many were seen as animals – Draconis, Leo, Pisces, Scorpius, Taurus . . . The ancients seemed eager to connect patterns in the sky with familiar forms of life on Earth.

Between the constellations can be seen many ghostly shapes: an elephant's trunk, horsehead nebula, dark clouds and striations of various forms that baffled astronomers at the turn of the twentieth century. Were these dark patches on the canvas of the night sky actual recesses in the distribution of stars, or were they foreground clouds of obscuring matter? There was a prolonged and heated debate on this question that continued for decades. Now we know that these dark patches represent dense clouds of cosmic gas and dust. They are the birthplace and nursery of stars. But to this subject we shall return in Chapter 4.

The sun, and indeed every other star in the Milky Way begins life as a fragment of a cosmic cloud, such as is seen for instance in the dust clouds in Sagittarius (see Plate 4.1 [top]). The composition of such a cloud is dominated by the lightest element, hydrogen, with 10 per cent by mass of helium and a smattering of heavier elements, including dust and molecules. The prestellar cloud contracts under the forces of its own gravity and heats up as it does so. The temperature at the centre of the contracting cloud continues to rise and the contraction is not halted until a central temperature of some 10 million degrees is reached. This initial contraction phase lasts on the average for about 50 million years.

Why do stars shine? When we light a fire, what happens is that the atoms of the ordinary chemicals in our fuel (coal or wood, for example) form new combinations with each other and in doing so release energy – heat. When the temperature at the centre of a star rises over 12 million degrees, the atomic nuclei begin to reshuffle their assemblage of protons and neutrons, and doing so liberate energy. A sequence of thermonuclear reactions then takes place in the central core of the fledgling star. The first set of nuclear reactions leads to the fusion of hydrogen into helium: four hydrogen nuclei (each comprising a single proton) come together at these high temperatures and unite to form a helium nucleus (composed of two protons and two neutrons). The minute mass difference that exists between a helium nucleus and four separate hydrogen nuclei is converted to energy according to Einstein's famous $E=mc^2$ equation which connects energy and mass. It is this energy release that provides the main power source of the Sun and indeed of most other stars for the major part of their lives. This source of nuclear power within stars collectively lights up the spectacle of the night sky.

In the case of the Sun, the highly stable phase of hydrogen to helium conversion is estimated to last for about 12 billion years; and now at some 5 billion years from the start, we are well settled at a point somewhat short of the middle of this phase. The end of the phase will be heralded when a core of helium amounting to about 12 per cent of the total mass of the Sun is formed at the centre. When four hydrogen nuclei combine to form one helium nucleus, the new arrangement reduces the pressure within the stellar core. This upsets the sensitive gravity–pressure balance that existed before and gravity squeezes in the core. The Sun then heats further in a shell around its core, nuclear reactions continue releasing more energy, and as result rapidly distends its outer layers, enlarging its size by a factor of 100–200. The surface temperature of the Sun drops from its present value of 6,000 degrees to below 4,000 degrees, and our parent star enters the red giant phase of its life. This stage is due to commence in another 5 billion years or so. Although the surface will get cooler, the increased surface area means that it will be about 100 times more luminous than it is now. The inner planets, Mercury, Venus, Earth and even Mars, will be slowly engulfed by the expanding atmosphere of the giant Sun.

Hydrogen burning continues for a little while more, now confined to a shell around the central helium core. Ultimately, the hydrogen fuel of the star will be effectively exhausted and this will be accompanied by further contraction and heating of the core. At the awesome temperature of 200 million degrees at the sun's centre, the helium is hot enough to undergo further nuclear reactions.

The question of how helium could be transformed into heavier elements opened a new chapter in astronomical history in the middle of the twentieth century. At first sight there appeared to be two possible ways by which this transformation could be achieved. The first involves the sequential addition of protons to the helium nucleus. The second involves the possibility of entire helium nuclei coming together to form heavier atomic nuclei, which means that the resultant atomic nuclei would each have multiples of two protons and two neutrons. Both these schemes for producing heavier chemical elements had apparently failed in the early 1950s – until Fred Hoyle argued that the rare encounters of three helium nuclei coming together at one place could produce nuclei of the element carbon at a fast enough rate, *provided* there existed an 'excited state' of the carbon nucleus, a state that had not yet been discovered. The carbon nucleus formed in such a hypothetical excited state could subsequently jump down to the lowest energy ground state, releasing energy in the process, and this energy release would keep the star shining for the next major phase of its life.

Because carbon is essential for life, Hoyle confidently predicted that such an excited state of the carbon nucleus must exist. Later experiments carried out by Caltech physicist William Fowler did indeed prove that this desired state exists, thus showing how one could understand the formation of the life-giving element carbon in the Universe, and the same time how the next important nuclear fuel – helium burning – switches on within stars. This nuclear reaction, known as the triple-alpha reaction, builds up the amount of carbon in the star's central core, while the amount of helium declines, until eventually this process too comes to a halt.

Once again the star's inner core contracts under the forces of gravity, the temperature rises further until the next energy-producing nuclear reaction can switch on. This is the reaction that unites a carbon nucleus, ^{12}C, with a helium nucleus, ^{4}He, to form oxygen, ^{16}O: another element that is vital for life. When that reaction too comes to a halt, the star enters its very last stages of the red giant phase. Its radius increases still further, vigorous convective motions are set up the outer layers, and processed material including carbon, nitrogen and oxygen boils off from the surface. In this way the atoms that serve as the feedstock of life are eventually injected into space.

Beyond the red giant phase the life history of a star turns out to be a trifle complicated, and the details depend strongly on how massive the star was in the first place. Stars in the mass range of twice the mass of the Sun to about half the solar mass (and that includes the Sun, of course) go through an early history broadly similar to that of the Sun. Such stars, in the final stage beyond the red giant phase, become the type known as a white dwarf. In such a star matter is in a highly compressed state, with particles jostling with each other and competing for the same space in a condition known by physicists as degenerate matter. A star with mass between twice and four times the solar mass also goes through a similar evolution to the Sun, only a little faster. It reaches the red giant phase in the manner we described above, loses a large fraction of its mass, down to 1.44 times the solar mass in the red giant phase in the form of stellar winds, and again ends up as a white dwarf.

Stars form from interstellar clouds in groups or associations. The most massive stars with masses exceeding ten solar masses will be the first to condense. They also evolve at the fastest rate, ending their relatively short lives as exploding stars known as supernovae. During the very last stages of their evolution nuclear reactions progress beyond carbon and oxygen to lead to the formation of iron and other metallic elements. At a critical point in a chain of nuclear reactions the massive star is no longer able to maintain its equilibrium and explodes.

The very last stages of nuclear processing in the explosion of a supernova were also worked out in the 1950s by Hoyle in collaboration with Fowler, Geoffrey and Margaret Burbidge. Supernova explosions represent some of the most energetic events in the Galaxy and demonstrate clearly how atoms of carbon, nitrogen, oxygen, phosphorous, iron and so on needed for life are returned into interstellar space. These atoms would then be available for inclusion in another generation of stars, planetary systems, organic molecules and life.

The most famous supernova in history occurred in the year 1054 and was recorded by Chinese astronomers who kept meticulous records of all celestial events. Chinese astronomers worked for the emperors and any errors of prediction or misreporting were considered heinous crimes often punishable by death. Of the 1054 supernova the record states:

> In the 5th year of the Ch-Ho regime, on the day of Chi-Chou, a guest star appeared within a few inches to the South-East of Thian Kuan. After one year it turned faint and disappeared.
> (from *The History of the Sung Dynasty* – Ho Pemg Yoke, 1962)

The record goes on to state that the guest star could be seen in daylight and that it was five times brighter than the planet Venus. There is no star to be seen today where the supernova once was, but long exposure photographs through a telescope reveal a filamentary nebulous blob on the sky known as the Crab Nebula. This is located some 6,000 light years away, and at the centre of the nebula reside the remnants of the exploded star, a pulsating star made wholly of neutrons known as a pulsar.

Astronomers estimate that there must be between two and three supernova explosions every century in the Galaxy or Milky Way system. But most of these will be so far away that their light is obscured by the clouds of absorbing dust that exist in the plane of the Milky Way.

The most thoroughly studied supernova in modern times occurred on 24 February 1987. This was a star that exploded not in our own galaxy but in a neighbouring galaxy known as the Large Megallanic Cloud (LMC for short). The star, which was identified after the event, was 15 times the radius of the Sun and 17 times its mass. It was a supergiant star with a surface temperature of 20,000 degrees and an intrinsic brightness equal to 40,000 times that of the Sun. Codenamed Supernova 1987A this object has given astronomers the best insight so far about the processes that occur in the final stages of a very massive star.

To sum up, there is now direct evidence that the essential feedstock of life, in the form of atoms such as carbon, oxygen, nitrogen, phosphorous and a host of metals, are expelled into interstellar space from stars in their red giant phases, and more dramatically from supernovae. In our galaxy alone a billion, billion, billion megatonnes of such material lies in wait to be taken up into new star and planet systems.

One of the most important questions that will be discussed in later chapters relates to how the life-giving chemical elements from stars come to be assembled into living structures in the cosmos. This is the question that asserts itself above all others whenever one ponders the mysteries of the Universe. And that question is intimately linked to another big question in science. How did the entire Universe begin? Our galaxy is the Milky Way, yet millions of other similar galaxies stretch out beyond the range of our most powerful telescopes. How far does it all extend? And how did this grand scheme of things come to be in the first place?

The high priests of modern cosmology claim to have an answer: it all began with a Big Bang and a singular creation event at a definite moment of time some 13–15 billion years ago. All the matter in the entire Universe was concentrated in a 'super atom' which exploded with unimaginable violence. In this Big Bang model of the Universe, everything is supposed to come into existence from nothing. The gigantic explosion of the super atom is thought to have given rise to the present-day expanding Universe, with each galaxy rushing away from every other with ever-increasing speed.

Apart from violating the normal principles of physics whereby matter and energy are conserved, Big-Bang cosmologies face very serious philosophical difficulties as well. What was there before the Big Bang event took place? What caused this event to occur? These questions are not permitted within the conventional framework. Big-Bang cosmologies have become fashionable because they are supposed to account for at least two sets of astronomical observations. The first is the detection of a background of microwave radiation, that was discovered by Arno Penzias and Robert Wilson in 1965, and that supposedly represents a relic of the energy released during the Big-Bang event. The second is the abundance of the light elements helium and deuterium which may have been formed soon after the Big Bang, when the temperature of matter exceeded tens of billions of degrees.

None of the available observations points uniquely to the correctness of the standard Big Bang cosmology, however. Some modern estimates of the age of the Universe, according to astronomical observations interpreted in terms of Big-Bang models, are turning out to be younger than the ages of the oldest stars in our own Galaxy. This, of course, is logically impossible: the Universe could not be younger than objects within it.

Astronomical data as they stand at the present time could fit equally well to a new class of cosmological model, 'Quasi Steady State Comology' (QSSC), that was proposed by Hoyle, Burbidge and Jayant Narlikar in 1993. In this class of model the Universe is supposed to have an infinite age and is infinite in its spatial extent as well. Over a timescale of 1,000 billion years (hundreds of times older than the ages of the oldest stars) the Universe expands. But over shorter timescales of about 50 billion years it oscillates, that is, it alternately expands and contracts. Creation of new matter takes place, not continuously and uniformly as was the case in an earlier version of Steady State Theory, originally proposed by Hoyle, Hermann Bondi and Thomas Gold, but only in the denser phases of the Universe and in the densest places like black holes. Many of the strongest philosophical objections to the standard Big-Bang cosmologies are overcome in the new model, although it is fair to say that the high priests of cosmology are yet to be convinced.

It remains to be seen whether the Quasi Steady State Cosmology of Hoyle, Burbidge and Narlikar holds up to observational verification and the ultimate test of time. But it has several points of appeal that relate to ideas we shall develop later in this book. If the origin of life turns out to be an insuperable problem in standard Big-Bang cosmologies, without the intervention of a Cosmic Intelligence, and if a mechanistic solution is desired, then QSSC could come to the rescue. The emergence of life could either require the cumulative effects of an infinite number of connected evolutionary steps in an essentially boundless Universe. Or alternatively, it never originated. It was always there as an ever-present attribute of the Universe.

References

HOYLE, FRED (1975). *Astronomy and Cosmology*, New York: W.H. Freeman.

HOYLE, FRED, BURBIDGE, GEOFFREY & NARLIKAR, JAYANT (2000). *A Different Approach to Cosmology*, London: Cambridge University Press.

NARLIKAR, JAYANT (1999). *Seven Wonders of the Cosmos*, Cambridge University Press.

WICKRAMASINGHE, N.C. (1975). *The Cosmic Laboratory*, University College Cardiff Press.

2. The Planetary Zoo

The mighty frame; how build, unbuild, contrive
To save appearance; how gird the sphere
With centric and eccentric scribbled o'er
Cycle and epicycle, orb in orb
John Milton, *Paradise Lost*, book viii *(1608–1674)*

Planetary systems are the dragons' dens, their breeding place. Here they sow their seeds, they grow, they kill. How did the solar system come into being? By understanding its genesis and history, could we know how planetary systems form? How many other similar systems exist? And what might be the prospects for life on alien extra-solar system planets?

Our solar system consists of the Sun at its centre surrounded by a number of planetary bodies mostly occupying the volume of a flattened disc. The planetary objects move in more or less circular orbits around the Sun, and they fall into distinct categories according to their distance from it. If the average distance of the Earth from the Sun is taken to be one astronomical unit, the Earth-like or so-called terrestrial inner planets – Mercury, Venus, Earth, Mars – occupy the region of the disc between one third and one and a half astronomical units. Between one and a half and five astronomical units there are the asteroids – a multitude of rocky bodies with sizes in the range 0.5–300 kilometres and with a total mass of somewhat less than a thousandth of the mass of the Earth. Then come the gas

giants Jupiter and Saturn at 5.2 and 9.5 astronomical units, next the
icy planets Uranus, Neptune and Pluto 19–40 astronomical units
away from the Sun. Next there is a flattened belt of 10–100 metre
sized icy comets, tens of millions of them, more or less in the same
plane as the planets, occupying a region 40–100 astronomical units
away. This is known as the Kuiper belt of comets. Further out there
is a gigantic shell of many hundreds of billions of comets, known as
the Oort cloud, surrounding the entire planetary zoo between radial
distances of 10,000–100,000 astronomical units from the Sun. A
host of Chinese dragons standing guard over the wealth of the
planets!

How did this weird system of objects come into being? The
origin of the solar system has been the subject of vigorous controversy for centuries. The French mathematician Pierre-Simon
Laplace (1749–1827), who lived in the so-called Age of Reason,
held the view that everything, from the motions of the stars to the
twinkling of an eye, could be explained by laws that governed
the Universe. Of course, all such laws in the Laplacian scheme of
things had to be devised by an all-powerful deity, who was,
predictably, a mathematician.

In 1796 Laplace adapted and modified an earlier theory of
the formation of the solar system attributed to the philosopher
Immanuel Kant. In this so-called nebular theory the Sun was
formed together with the planets from a single rotating gaseous
nebula. The sun contracts from the inner hotter regions, while the
planets condense out from the colder outer parts of a flattened disc-
shaped nebula. Although the Kant–Laplace model had many critics
in the early days, some form of this theory is now thought to be
correct. In most modern versions of the theory the entire solar
system is considered to have formed from a single cloud of
interstellar matter.

Over 5 billion years ago all the atomic material making up the
solar system, the Sun, planets, satellites, comets and life would
have been contained within a small part of a very much larger cloud

of cosmic gas and dust. Such clouds are all-pervasive within our Milky Way system and are recognised by astronomers as 'giant molecular clouds'. Within these clouds all the atoms needed to make planets are to be found – H, He, C, N, O, Fe, Mg, Si – as indeed are a host of molecules that may be connected with life.

Stars do not form singly, but in very large numbers, and the Sun would not have been an exception to this general rule. The molecular cloud from which the Sun formed, as one of a multitude of stars, is believed to have torn itself apart from the rest of the galactic gas, and begun to collapse under the forces of its own gravity, nearly 5 billion years ago. In the process of falling together the cloud would have become progressively denser and started to flatten somewhat. Fluctuations of temperature and gas pressure occurring naturally within the cloud would have inevitably led to the cloud becoming clumpy.

Slightly denser blobs of gas and dust would arise hither and thither in the cloud, and these would have led to independently contracting cloud fragments. There would have been a spread of mass within these clumps or cloud fragments. The masses of individual fragments may have ranged from a fraction of the mass of the Sun, to tens of times its mass. Each of the fragments was to become a star. The largest of the stars to form in this ensemble would have evolved at the fastest rates (see Chapter 1). They would go through the successive steps of nuclear processing – hydrogen burning, helium burning, carbon burning and so forth, until eventually reaching the supernova stage. The explosions of such supernovae would have scattered a rich harvest of atomic nuclei needed for life, that would then have been available for other star systems, including the solar system, as they formed.

There is evidence to suggest that the particular fragment of the molecular cloud that eventually formed our solar system was slow to get started, until a supernova had exploded in its vicinity. The evidence includes the discovery of tiny particles of diamond and also of certain radioactive elements within meteorites, that could only have come from a supernova explosion. Whether there was just one or several supernova explosions associated with the early history of the solar system is a matter of dispute at the present time.

A supernova explosion would surely have led to a shocked gaseous envelope travelling outwards into the surrounding parts of the fragmented molecular cloud. Shock waves could thus have triggered the ultimate collapse of the cloudlet that went to form our solar system. Most of the material of the cloudlet may have collapsed into the central star which is the Sun. In the Hoyle–Kant–Laplace model of planet formation the laws of physics dealing with rotation and magnetic fields took care of the rest.

Rotation or spin is a common feature of astronomical bodies. The entire galaxy, the disk-shaped volume that is filled with the stars of the Milky Way, spins around a central axis, completing a full revolution in about 250 million years. Individual clouds of gas and dust which are interspersed between the stars have their own random motions in the galactic disk as well as their own rotations. Stars too have their own spins. The Sun, which is our private star in the solar system, turns about its axis once in 26 days, and the Earth turns about its north-south axis once in every 24 hours.

The Sun's slowness of spin might provide a crucial clue as to how the planets, including the Earth, came into existence. If the Sun was formed by the contraction of a cloud of spinning inter-stellar dust, it is possible to show that it should be rotating many times faster than it now does. In fact it should rotate about its axis once in a fraction of a day, not once in 26 days. This is a con-sequence of the laws of motion, first formulated by Isaac Newton (1643–1727), which states that the rotational momentum of any mass of material which is not acted on by external forces cannot change. The consequence is that when the same amount of matter becomes compressed into a smaller object, the compressed body must spin faster in order to contain all the rotational momentum of the dispersed object. The situation being similar to what happens

19

when a figure skater on ice draws her arms close in to her body and thereby spins faster.

What then became of this rotational momentum in the case of the Sun? The most convincing answer was given by Hoyle many years ago. If one were to add up all the rotational momenta of all the planets in the present-day solar system, taking due account of the hydrogen and helium that was initially associated with the planetary material, it turns out that we have a full account of the missing rotational momentum of the Sun. It appears then that in some way the contracting primitive Sun had given up most of its original rotational momentum to the much smaller total mass associated with the planets.

There is a very simple way in which this transference could have been achieved. As the cloud of gas and dust that went to form the Sun contracted, it began by conserving all the rotational momentum for the reasons just given. As the cloud became denser and smaller it would rotate faster, and as the contraction proceeded, the entire structure would flatten near the poles in the manner of a thick spinning ball of dough. A crucial stage will have been reached at some point in this process when the gravitational forces in the primitive Sun would no longer have been able hold up against the strength of the rotary forces that developed, particularly near the flattened equatorial regions. A disc of material would have been flung out from the equator, and it is from this disc that the planets and comets were eventually born.

The gaseous material escaping from the Sun would have been partly split into nuclei and electrons (partially ionised), and in such a gas magnetic lines of force would anchor the gas to the parent sun, like spokes in the hub of a bicycle wheel. It is this magnetic anchoring that could have caused the transference of rotational momentum from the Sun to the planetary gas, thus resulting in the slowing down of the Sun's rotation. This process of magnetic braking of the primitive Sun was proposed independently by Hannes Alfvén and Hoyle. The speed of ejection of the gaseous material as it left the Sun would have been sufficiently great for the planetary disc to get flung out with enough energy to reach the distances of the outermost planets Uranus and Neptune. From these outer regions of our solar system the lightest gases hydrogen and helium in the planetary disk would have escaped.

The chemical history of planetary material from the moment it left the equator of the fledgling Sun to the various stages of planet condensation can be calculated from the known physical and chemical properties of the various substances that are involved. Such calculations were first done by Hoyle and myself in the late 1960s. When the planetary matter left the Sun its temperature is estimated to have been about 3,000 degrees. At such a temperature the escaping gas would have been in the form of single atoms of hydrogen, helium, oxygen, carbon, nitrogen, silicon, magnesium, iron, sulphur, phosphorous and so on, atoms that were synthesized at an earlier time in the interiors of stars. In the course of its journey outwards, the temperature, density and chemical composition of the disk all change in a manner that can be calculated precisely from known theories of physics and chemistry. As the disk expands away from the Sun, it cools and we can show that various types of molecules form. While the tightest bound molecules and solids form closest to the Sun, and the weaker bound molecules and solids form in the colder outer parts of the disc.

Our first numerical computations on such condensation sequences were carried out in the 1960s using the computer at Cambridge University known as EDSAC2, a machine that had less computing power in a whole building than we now have in a powerful Pentium PC. The calculations showed that the first solid particles to condense from an expanding gaseous disc were composed of iron and silicates, similar to the material present in the Earth's core and crust. This happened when the gas had cooled to 1,500 degrees, about the temperature of a blast furnace. We found it significant that such materials do indeed condense and fall out of the gas at just the distances of the so-called terrestrial planets – Mercury, Venus,

Earth and Mars. When they first condense from the gas the iron/ mineral dust particles are nearly molten and therefore tend to be sticky. They start off with sizes smaller than pinheads, but as they collide with each other they stick and clump together to form larger and larger chunks. When, by a process of 'snowballing', the chunks reach sizes greater than about a metre, they fall out into circular orbits around the Sun at exactly the places where they condensed. Such clumps then become too heavy to be carried further out by the drag of the expanding planetary gas.

To begin with, the region of the present-day terrestrial planets would have been strewn with metre-sized débris, billions of pieces orbiting around the Sun. Occasionally these pieces, still quite sticky, would collide to form a slightly bigger chunk. From this point on planet formation is a little like a winner-take-all race. The larger chunks grow faster than the smaller ones, analogous to snowballs rolling down a ski slope, the bigger they get the faster they grow: such chunks of débris heading towards a planetary condition are called planetissimals. The crucial dimension for a growing planetissimal to become a planet is about a kilometre, because objects this size begin to pull in neighbouring chunks by means of their own gravitational pull. This general scheme of events could account for the formation of Mercury, Venus, Earth and Mars. In each case a well separated large object of kilometre size began mopping up débris that lay in its path. Calculations show that the formation of the terrestrial planets took a relatively short time, perhaps only a few million years. Planetissimal aggregates that failed to make it beyond the critical size to grow into planets would have inevitably populated an orbital belt just outside the terrestrial planets that we can now identify as the asteroid belt. Asteroids on this picture are therefore the pieces of a planet or planets that never came into being.

A triumph of our particular theory of planet formation is that we can immediately understand why the Earth and inner planets are made mostly of iron and rock, and why they are found at their present distances from the Sun. These features were simply determined by the conditions that prevailed at the surface of the primeval Sun at the time when the planetary material left it, and by the well-attested laws of chemistry and physics that determined a condensation sequence of solids. In this scheme of things volatile substances like water and carbon dioxide, so important for life, could only have been trapped in rather small quantities as water of hydration in rocky material and in carbonates that formed the outer crusts of the terrestrial planets and would have been quickly lost. Further additions of these volatiles that were needed to form the Earth's oceans and atmospheres must have come from the outer regions of the solar system at a later date, from comets, as we shall now see.

The planetary gases that left the Sun did not all stop at the places where the inner planets and the asteroids had formed. The uncondensed smaller chunks of iron and silicates and the bulk of the hydrogen, helium, carbon, oxygen, and nitrogen expands further outwards from the sun like an army in retreat. Eventually the expanding disc would have reached the present-day orbits of the planets Jupiter and Saturn, and thence Uranus, Neptune and Pluto. At the distances of the orbits of the giant planets Jupiter and Saturn, some of the rocky bolides that were not mopped up into the terrestrial planets may have coalesced and fallen out into orbits, where mutual collisions would first have led to the formation of iron or mineral cores, with later accretion of hydrogen and helium (100 Earth-masses each) to form the gas giants Jupiter and Saturn.

It is to the formation of the outer planets Uranus and Neptune that we must turn in order to understand the origin of comets and their effect on the Earth. When the expanding planetary gases reached the orbits of Uranus and Neptune the temperature of the gaseous disk would have been just correct for molecules like water and carbon dioxide to form in the gas and for the condensation of ices to occur. The ices, dominated by water ice, would soon form themselves into considerable sized chunks, and there would be very many such chunks – about 50,000 billion billion billion individual chunks, at a rough estimate, each measuring a couple of metres across. These icy chunks with their enormous total surface area would have been able to mop up any residue of dust from interstellar space that was left behind at these great distances in the fragment of interstellar cloud that went to form the solar system.

Within a mere 1,000 years or so (a few orbital periods) the icy chunks would have collided and settled into very nearly circular orbits, occupying a thin band of orbits. The next step would involve a process of aggregation of icy bolides laden with interstellar dust within this band, leading to the formation of compact objects of radii in the range of 50–200 kilometres. Such objects might be identified with the class of giant comets that was originally predicted by Victor Clube and Bill Napier, of which about 100 have so far been identified. The final step in the entire planet-forming process would appear to have been the aggregation through collisions of these giant comets to form the icy planets Uranus and Neptune. This last phase of planet formation is estimated to have taken about 300 million years.

Not all the colliding icy planetissimals went to form the planets Uranus and Neptune, however. A fraction would have remained to define a belt of comets just outside the orbits of Uranus and Neptune, 30–100 astronomical units from the Sun. The existence of such a belt of primordial cometary objects was first suggested by Gerald P. Kuiper as early as 1950, and the object Chiron was recognised as being an object of this type in 1989.

A mass of cometary objects of radii in the range of 10–100 kilometres, comparable in total mass to the combined masses of Uranus and Neptune, would have been thrown out by gravitational interactions into orbits that took them out to great distances from the Sun. It is this latter process that is believed to have led to the formation of the present-day reservoir of comets known as the Oort cloud.

There is a further consequence of the formation of Uranus and Neptune and the generation of the Oort cloud of comets that is of particular relevance to the Earth. A fraction of the icy comets in the outer solar system would have, through dynamic interactions, been flung into orbits that plunged into the inner regions of the solar system. This process resulted in what is known as the late accretion epoch of the Earth, which involved a protracted period of cometary impacts. We shall see in Chapter 3 that the Earth's oceans, atmosphere and even life may have been brought in this way.

We have so far made no reference to the many satellites or moons that orbit the planets. These satellites vary in size, composition and geological history, and have recently been the focus of several programs of space exploration. The Earth's own Moon is among the larger of the satellites in the solar system with a diameter of a little more than a quarter that of the Earth. Phobos and Deimos, the satellites of Mars are among the smallest, measuring a mere 14 and 8 km across. The elaborate satellite systems of Jupiter, Saturn and Uranus each resemble miniature solar systems to varying degrees.

There are two main processes by which planetary satellites may have been formed. One process involves capture of large planetissimals by a gravitational capture process. The other involves an impact by a large object that vaporises the outer layers of the planets. The vapour is thrown into a gravitationally bound envelope from which a satellite or satellites could form, following essentially the same chain of events as we have described for the formation of planets around the Sun. Once formed, the larger satellites will evolve geologically as a result of internal heat sources, tidal interaction with the parent planet as well as cratering impacts by comets and asteroids. Recent studies of the surface of Europa, one of the four main satellites of Jupiter, have revealed tantalising signs of a subsurface ocean underneath a frozen icy crust, and possibly life.

According to the foregoing discussion the development of a planetary system has to be considered as being a natural consequence of the formation of a sunlike star. Planets may therefore be expected to orbit a good fraction of sunlike stars in the galaxy.

Where then, one might ask, is the evidence for such planets? The search for extra-solar planets and planetary discs has only just started, and already several remarkable successes have been reported. Planetary or protoplanetary discs have been detected by their infrared or heat emissions around a number of young stars, the disc around the star Beta Pictoris being perhaps the most spectacular. The direct telescopic detection of planets is difficult for the simple reason that the light reflected by a planet from its central star is too faint to observe with present-day techniques at normal stellar distances.

On 6 October 1995 Swiss astronomers Michel Mayor and Didier Queloz discovered the first planet outside the solar system. This particular planet circles the sunlike star 51 Pegasi, which is at a distance of 50 light years from the Sun in the constellation of Pegasus. The planet was not seen directly by its reflected light but detected by observing a minute wobbling of the central star caused by the planet's gravitational pull as it orbited the star. It was not a planet like the Earth, but more like Jupiter, and the further surprise was that its orbit hugged close to the central star – at just one twentieth of the orbital distance of the Earth from the Sun. The technique of detecting planets by such wobbles selects only those planets that are rather massive like Jupiter, and also those that orbit close enough to the central star to produce detectable wobbles. To date, surveys of stars of types similar to the Sun have resulted in the detection of more than 32 extrasolar planets. Most of these have masses similar to Jupiter and they hug too close to their parent stars for life to survive on them or on any satellites or moons they may have. This is all that the planet hunt up to 2001 has so far revealed. The vast majority of life friendly Earth-like planets is still lying out there waiting to be discovered by a new generation of telescopes and techniques.

References

HOYLE, FRED (1978). *The Cosmogony of the Solar System*, University College Cardiff Press.

MARCY, G.W. & BUTLER, R.P. (2000). *Publications of the Astronomical Society of the Pacific*. 112, 137.

MAYOR, M. & QUELOZ, D. (1995). *Nature*. 378, 355.

NARLIKAR, JAYANT V. (1999). *Seven Wonders of the Cosmos*, Cambridge University Press.

SEEDS, MICHAEL A. (2001). *Astronomy: The Solar System and Beyond*, Brooks/Cole.

3. A Planet Grooms
for Life

The heavens limitless
Earth floating through the void.
Not a conception, not a fancy
But actually existing . . .
Daisaku Ikeda *(1928–)*

The third planet from the Sun is our home, the Earth. Its place in the wider world and its part in the grand scheme of things has been the subject of discussion and enquiry for millennia. In the sixth century BC Pythagoras is credited to have declared that the Earth is a sphere resting at the centre of a spherical Universe. Earth-centred cosmologies in a variety of forms continued to dominate Western thought for centuries, running through the philosophies of Plato and Aristotle and culminating in the work of Ptolemy. To account for the apparent motions of the Sun, Moon and planets, as they were known around the second century AD, Claudius Ptolemaeus (Ptolemy) devised an ingenious geometrical scheme of epicycles which he published in an encyclopaedic work of 13 volumes under the title *Almagest*. This was undoubtedly one of the most important sets of books ever written and it had a profound influence on the philosophical outlook of many generations that were to follow. The Ptolemaic world view was fully in keeping with earlier Pythagorean

28

cosmologies that enshrined the principles of perfection and symmetry. As new and more refined planetary data became available, particularly after the advent of telescopes, the Ptolemaic scheme had to be modified considerably by adding new and more complicated epicycles. By the middle of the sixteenth century the system of epicycles that was required to explain all the astronomical observations had become so cumbersome that it eventually had to be rejected in favour of the Copernican Sun-centred world view.

There were of course sages of old who espoused post-Copernican world views long before the dawn of the sixteenth century, but their ideas failed to reach the attention of mainstream philosophers. The *Vedas* and *Rigvedas* of ancient India described cosmologies that were distinctly non-Earth-centred as early as 1500BC. In the West Aristarchus of Samos (*c*. 270BC) boldly placed the Sun amongst the stars and held that the Earth revolved round the Sun. Aristarchus's heliocentric theory was evidently unknown to Nicolaus Copernicus when he wrote his classic *De revolutionibus orbium coelestium libri VI* eighteen centuries later in 1543.

The monumental work of Copernicus led eventually to a firm rejection of earlier Earth-centred cosmologies with far-reaching implications for science and society as a whole. The Earth was at long last dethroned from its hitherto privileged position at the centre of the Universe. And further humiliations followed with the expansion of astronomical knowledge in the twentieth century. Our planet, our solar system, our Galaxy have all faded away total insignificance as astronomers probed the Universe further and deeper than ever before. The Earth was relegated to the status of just one of nine planets orbiting the Sun, the Sun in turn to just one of a 100 billion similar stars which form the Milky Way, and our Milky Way itself one of hundreds of billions of galaxies that make up the observable Universe. We could no longer justify any claim of special status for the Third Planet nor indeed for the life that developed upon it.

In Chapter 2 we saw how the formation of the Earth was connected to the formation of the Sun over 5 billion years ago. The fragment of a giant cloud of interstellar gas and dust from which a clutch of new stars were being formed was struck by a powerful shockwave, an explosive blast that marked the death of a massive star – a supernova. A slowly-rotating solar nebula was thus born. The far-flung atoms of the nebula, mostly hydrogen and helium with traces of heavier elements from earlier stellar deaths, were compressed through the action of gravity to form the central star, the Sun. We saw earlier how the shrinking proto-sun spins progressively faster as figure skaters do when they draw their arms close into their bodies. A stage comes when the fast rotating star squirts out a gaseous disc from its equatorial regions, from which proto-planets and finally the terrestrial planets, including the Earth, condensed.

As rocky asteroidal objects fell together to form the primitive Earth, some of the gravitational energy that was released melted the interior of the planet as it grew. The molten rock was then partly separated according to density, the denser minerals sinking deeper, the lighter rocks rising to the surface. At the start of its history as a solid body 4.5 billion years ago, the Earth was already partly differentiated in this way. When the surface had cooled to the temperature of the surrounding space the planet's crust would have cracked and shrunk. The shrinking surface layers would have compressed minerals below, thus continuing to heat the planet's interior.

The heating at the Earth's centre would have been greatly enhanced by the presence of radioactive elements, uranium and thorium, that were synthesised in an exploding massive star. As these unstable elements decayed they would have released high-energy particles and radiation which would have heated and liquefied the rock deep below the crust. This process would have led to a major restructuring of the planet's interior on a relatively short timescale. Gravity acting on the heavy elements iron and nickel would have made these materials sink to the central core, whilst the less dense silicates would have floated to form the mantle. All this would have happened relatively quickly, within the first 100 million years or so of the Earth's life.

Whilst the Earth's mantle and core were being restructured, the exterior crust was also being remoulded with a vengeance through external impacts. When the outer planets Uranus and Neptune were still being formed from collisions between icy comets, some of these comets would have been deflected into the inner region of the solar system and crashed on to the surface of the terrestrial planets, including the Earth. Such collisions, taking place over tens of millions of years would have added volatiles such as water and carbon dioxide to the Earth's surface. Comets are thought nowadays to have been the major source of water in the oceans. The evaporation of water from the oceans and the break-up of water molecules by solar ultraviolet light then led to an atmosphere and a cloud cover around the planet.

Evolution and remoulding of the Earth's surface has continued unabated for the past 4.5 billion years. Molten lava, that is present deep in the mantle, wells up through fissures in the crust from time to time, possibly assisted by impacts. Sections of the crust slide over each other, pushing up mountain ranges and shifting continents. Comet and asteroid impacts produce craters, such as Meteor Crater in Arizona. Surface irregularities caused by whatever processes do not however persist for indefinite lengths of time. Wind, water and sliding glaciers more or less continually erode the surface, smoothing out irregular geological features.

Until recently scientists believed that the frequency of comet impacts would have prevented the formation of a stable crust and an atmosphere on the Earth until about 4 billion years ago. Studies of lunar craters show that both the Earth and Moon were subject to extremely heavy bombardment up to precisely this moment of time. There are also recent arguments to suggest that about 4.45 billion years ago, just 50 million years after the Earth formed as a solid body, a Mars-sized body slammed on to it, blasting forth a spray of orbiting particles from which the Moon coalesced. This theory has been put under strain by another recent discovery, however. In 2001 Stephen I. Mojzsis and his colleagues discovered a 4.3 billion-year-old zircon crystal in a rock, which they argued must have interacted with cold water at the Earth's surface. The implication is that substantial domains of water must have existed at the surface scarcely 150 million years after the postulated Moon-forming event, and still well within an epoch of intense bombardment of both the fledgling Moon and Earth, the Hadean epoch as it is called.

Whilst the oldest evidence of a stable crust and liquid water at the Earth's surface may possibly be dated at 4.3 billion years, the oldest biochemical evidence for life on the Earth is dated at 3.8 billion years. The latter evidence is derived from a subtle measurement of the distribution of carbon isotopes in ancient carbonate deposits. A few years ago in 1996, a team of scientists also led by Stephen J. Mojzsis showed that in such ancient rocks there was a slight fractionation of the lighter isotope of carbon, ^{12}C relative to ^{13}C. This is just what one would expect from biochemical processes associated with life that slightly favour the lighter carbon isotope. Recognisable shapes of bacterial fossils, on the other hand, appear at a slightly later date in the geological record, and according to recent studies by William Schopf this evidence is found in rocks with ages of about 3.5 billion years.

The violence that marked the beginnings of the Earth did not entirely stop at the moment when the first life appears. There is evidence of sporadic impact events punctuating both the physical and biological history of the planet all the way down to the most recent times. These matters will be the subject of more detailed discussion in Chapter 8, but for the moment let us note that the third planet has indeed witnessed many episodes of violence and adversity in the past, some at least being dictated by impacts. Its climate has oscillated wildly between extreme glacial conditions and much warmer interglacials, and life too has intermittently suffered episodes of extinction as well as huge bursts of speciasation. The extinction of the dinosaurs 65 million years ago is an example of the former and the Cambrian explosion of species 570 million years ago is an example of the latter. There is evidence to support the view that immediately prior to the Cambrian explosion of life there may have been a near-total extermination of all prevailing life, a fully glaciated 'Snowball Earth' that lasted for nearly 10 million years (see Chapter 11). It seems likely that events of this type were all caused by the interaction of the Earth with comets.

Against the background of such a history of impacts it is easy to see how the development of terrestrial life from single-celled archaebacteria (archea) to highly complex and varied life forms was a slow and chancy process. The evolution of life had to take place whilst the continents of our planet and ocean floors were on the move. The ocean floors are recycled into the Earth's interior on timescales of hundreds of millions of years, whilst the continents that contain the oldest rocks drift apart at a very slow rate.

The continents have been on the move ever since they broke apart from a single supercontinent 180–200 million years ago, a continent which has been called Pangaea. Pangaea was the brain-child of the German meteorologist Alfred Wegener who published his theory of continental drift in 1912. Wegener's ideas were not immediately received with favour, but now they have come to be widely accepted without dissent. The most conclusive evidence for this process comes from studies of the magnetism of rocks. Such measurements show how continents have changed their relative positions over millions of years, and moreover enable projections to be made into the future. How the original continent of Pangaea came to be split is still uncertain, but impact of comets could well have been a contributory cause.

Our comprehension of the passage of geological time is of course subjective, and our best understanding of time must relate to our personal experience. In the words of Sir Thomas Browne (1605–82):

Who can speak of eternity without a solecism, or think thereof without an ecstasy? Time we may comprehend, 'tis but five days elder than ourselves . . .

Religio Medici, pt.i, sec.1

As members of a species with an average life span of perhaps fourscore years we might best comprehend vast stretches of geological time by relating such times to the duration of human generations. Through contact with our parents or grandparents we might expect to feel a more-or-less direct contact with events spanning three human generations, perhaps 100 years. Beyond that our direct experience runs out and we must appeal to our intellect to attempt to understand the past. It is therefore not too surprising that human beings had no inking of the enormous antiquity of our planet or of life upon it until relatively recently. As late as 1650 Archbishop James Ussher relied on The First Book of Genesis and on the genealogy set out in the Old Testament to propose that the Earth was created 9 a.m. on 23 October 4004BC, and Noah's flood happened 1,655 years later in the year 2349BC.

Now we know better. The fossil record faithfully traces the progress of life over the past 4 billion years of geological history. Radioactive dating techniques also date many of these fossils with a remarkable degree of accuracy, so an evolutionary history of life is literally cast in stone. The fossils of the most complex life forms, plants and animals, occur in relatively recent times. The earliest fossils of *Homo sapiens* go back only a few 100,000 years, the earliest fossils of *Hominids* to 4 million years, mammals to about 150 million years, insects to about 300 million years, and the beginnings of metazoa life to 570 million years.

To put this record of terrestrial life into an easily comprehensible form let us scale down the well attested 4,500 million-year age of the planet so as to occupy a mere century. The following calendar of events then presents itself as a rough presentation of modern geological data. On this compressed scale of time the Earth is formed at the beginning of the century, exactly 100 years ago. The Moon-forming event is dated at 98.9 years ago, and the age of the first zircon crystals showing contact with masses of cold water at the Earth's surface is 95.6 years. The first isotopic evidence of microbial life on the Earth now shows up at about 85 years ago, and this also coincides with the time when comet impacts on the Earth had died down to a trickle. The famous Burgess Shale and Cambrian explosion of multicellular life occurs 13.1 years ago. Dinosaurs emerge 5.5 years ago and they become extinct less than a year and a half ago. Our direct line of descent, *Homo erectus*, a hominid walking upright on two legs made its first tentative appearance three weeks ago. And our immediate ancestor *Homo Sapiens*, hunter and food-gatherer, with brain equipped eventually to write the plays of Shakespeare as well as to unlock the secrets of the Universe came in as late as eight hours ago. Eight hours in a century-long span of terrestrial life: that is the sum total of our proprietory claim of this planet, no more, no less. The ancient Indus Valley civilisation and the city of Mohenjo-daro is scarcely an hour old; the nuclear era only seconds old; and the space age is virtually new born.

Enough of our Lilliputian indulgence. Let's return to the real world and to real uncompressed time and recall that at the start of our planet's history 4.5 billion years ago conditions at the surface were so hostile that no life – not even organic molecules – could have survived. Then suddenly, almost as soon as the epoch of intense cometary bombardment came to an end, the first signs of life appear. How did the transformation from inorganic matter to life occur? In the rest of this chapter I shall trace the progress that has been made towards answering this question, which is perhaps the most important question in the whole of science. Was there ever such a transformation that took place on Earth? Or was life brought to Earth along with the impacting comets?

In the long-gone days of *vitalism* people had the comforting belief that the substances that make up life were somehow different from the inorganic components of nonliving matter. This ancient conviction that dates back to classical Greece could only have been strengthened if vitalists had in their possession the modern armoury of molecular biology. When the genetic complexity of life came to be fully recognised in the middle of the last century, it became obvious that the information content of life, which gave life its special quality, had of necessity to be of cosmic order. At a molecular level life even in its primitive form is so incredibly complex that any prospect of transforming an inorganic system into biology must be considered awesomely difficult to say the least. Although the vitalists were wrong in the details of their beliefs, it would appear that they had a correct instinctive perception of the astronomical order of complexity inherent in life.

Vitalism was decisively rejected at the dawn of the nineteenth century after Friedrich Wöhler had shown how urea could be produced from ammonium cyanate. The belief then grew that the emergence of life from nonlife must involve processes that could not be different from ordinary chemical reactions that occur in a chemist's laboratory. This gave a fresh boost to another ancient doctrine, that of spontaneous generation, according to which life was supposed to arise naturally from nonliving matter under suitable conditions. In the third century BC Aristotle pronounced that fireflies emerged from a mixture of warm earth and morning dew, but in the nineteenth century the idea of spontaneous generation was applied to the less dramatic case of microscopic life forms, manifesting itself, for instance, in the souring of milk and the fermentation of wine.

Then spontaneous generation was also proved wrong. In 1857 the classic experiments of Louis Pasteur showed clearly that micro-organisms are always derived from pre-existing microbes, just as in the case of fireflies, larger creatures, plants and animals alike. So the clear logical deduction is that life as we know it here on the Earth is *always* derived from life that existed before.

The life-from-life relationship discovered by Pasteur permits a much wider generalization, which was expressed succinctly by the physicist Hermann von Helmhotz in 1874 in the following terms:

> It appears to me to be fully correct scientific procedure, if all our attempts fail to cause the production of organisms from non-living matter, to raise the question whether life has ever arisen, whether it is just as old as matter itself, and whether seeds have not been carried from one planet to another and have developed everywhere where they have fallen on fertile soil . . .

Other contemporary physicists, notably Lord Kelvin and John Tyndall, held similar views. Sir William Thomson (Lord Kelvin) said of Pasteur's paradigm:

> Dead matter cannot become living without coming under the influence of matter previously alive. This seems to me as sure a teaching of science as the law of gravitation . . .

On 21 January 1890 Tyndall gave a Friday evening discourse at the Royal Institution in London and by means of a simple optical experiment was able to show the presence of organic dust in the atmosphere. He then went on to make the bold assertion that the dust could include what he called 'vibriones' – germs that were carried in the air and caused epidemics of infectious disease. The editorial columns of the newly founded *Nature* magazine were quick to condemn him for reviving an ancient doctrine that was known by the name *panspermia*. Panspermia, implying that the seeds of life are inherent in the Universe, is an idea of great antiquity. It was implied throughout ancient Vedic and Buddhist cosmologies and in classical Greece the philosopher Anaxoragas also imagined that life infused and spread itself throughout the cosmos.

The first decisive steps towards reviving panspermia in modern times were taken by the Swedish chemist Svante Arrhenius

(1859–1927). In *Worlds in the Making* (1908), Arrhenius proposed an explicit mechanism by which bacterial spores could be transferred from one star system to another, noticing that the sizes of bacteria are related to the wavelengths of radiation from dwarf stars in a way that would permit the pressure of starlight to propel them across galactic distances. He had also correctly appreciated the exceedingly robust nature of bacteria and the survival properties of bacterial spores.

We have already referred to the connected sequence of fossils and microfossils in geological strata of different ages, going back to the moment when the first unequivocal signs of terrestrial microbial life are found. The time available on the Earth for the gentle brewing of a primordial soup leading to life has now all but vanished from the geological record. This moment of time is currently thought to be about 3.8 billion years ago, which already puts the first life on the Earth in the harshest of nurseries. But the very first evidence of life could well turn out to be even earlier, perhaps running into the epoch of intense cometary collisions – the late accretion epoch as it is called – but after the Moon-forming event at 4.45 billion years ago. Whenever this moment finally turns out to be, the situation before the appearance of the first life leads logically to two distinct possibilities.

(a) There is a requirement for some form of spontaneous generation of life on the Earth in a vanishingly thin slice of geological time – thereby denying the results of Pasteur, or

(b) There was no spontaneous generation on the Earth and the life-from-life connection is maintained by means of the importation of life from outside.

Attempts to create life from nonliving matter did not stop in the nineteenth century with the negative results of Pasteur's experiments. In the 1950s ideas akin to the ancient doctrine of spontaneous generation staged a dramatic comeback. The emphasis shifted now to making the building blocks of life from inorganic chemicals under laboratory conditions, conditions that were thought to mimic those that prevailed on a primitive Earth. At the time the prevailing view was that the primordial Earth's atmosphere was a mixture of partially reduced gases dominated by H_2O, CH_4 and NH_3. From such a mixture placed in a laboratory flask Stanley Miller, Harold Urey and their collaborators were able to show that trace quantities of life's building blocks can be formed through the action of electric discharges and ultraviolet light. The original molecules in this mixture are burst into fragments (radicals) and in the ensuing cascade of re-combination reactions trace quantities of organic chemicals are formed. The products included amino acids, which are the building blocks of proteins, and nucleotide bases and sugars which are the components of DNA and RNA. The results of these experiments were hailed as a major breakthrough in our understanding of life's origins, and it was optimistically thought that it would be only a matter of time before the spontaneous generation of life from inorganic matter would finally be achieved.

Despite a sustained effort on the part of many scientists the world over that final goal has become ever more remote. The most difficult problem of all is to explain transition from building blocks to edifice, from relatively simple chemicals to the incomparably magnificent edifice of life. That requires the invention of a genetic blueprint embodying a catalogue of building instructions, information that is specific in kind and unimaginably vast in quantity. Attempts to quantify this information, for instance by the number of random trials required to achieve the correct arrangements of building blocks for some crucial life molecules, gives rise to numbers that can only be described as superastronomical.

Would random shuffling in a warm little terrestrial pond filled with all the right biochemical monomers – amino acids, nucleotides, sugars – lead to the specific molecular arrangements found in biological structures such as DNA and RNA, or in the enzymes for which these structures code? A typical enzyme is a chain with about 300 links, each link being an amino acid, of which there are

20 different types used in biology. Biochemists studying the structure of several genes have shown that about one third of the links must have an explicit amino acid from the 20 possibilities. This means that with a supply of all the amino acids supposedly given, the probability of a random linking of 300 of them to produce a particular enzyme is as little as 1 in 10^{250}. And with 2,000 enzymes in a typical bacterium the odds work out at about one part in $10^{500,000}$. Some recent work by A.S. Mushegian and E.V. Koonin shows that the number of enzymes in the most primitive cells could be reduced to about 256. The chances are then a little improved, but still woefully inadequately: one in some $10^{33,300}$.

There can be scarcely any argument that the odds *against* the emergence of life in this way are superastronomical. Attempts to cut down these odds have, in our view, involved 'fudge' factors, but even so one always ends up with numbers that are in the same class of superastronomical numbers.

Self-organising properties of chemical systems that could *conceivably* lead to life have been discussed for some years, but here again identifying any such prospective process is proving more difficult than anyone might have imagined. Attempts to understand an origin of life on the Earth have recently focused on a hypothetical RNA world. The argument is that RNA could hold some amount of genetic information as well as reproduce it without the need for enzymes, DNA or any cellular machinery. The supposition is that the origin of life via RNA involved progression along a sequence of very many steps, each individual step being unrelated to life and requiring a vastly reduced level of improbability than the final hurdle to be overcome. It is like a blindfolded mountaineer climbing an infinitely high and infinitely wide range of mountains, taking infinitesimally small upward steps. Could such a process ever lead the mountaineer to a predetermined peak – one particular desired peak from amongst an infinity of equally probable peaks that might be reached? An affirmative answer implies that one could overcome impossible odds, or that there is an yet undiscovered guiding force that takes one to the desired summit. In the biological case a principle of determinism is conceded, the implication being that the final highly specific information content of life is somehow contained in the laws of physics. Such an assumption has no empirical basis and so the idea has to be considered as faulty, and possibly even lying outside the purview of empirical science.

The theory of origins based upon an RNA world is also facing a more direct challenge. Carl Woesse, the architect of the 'tree of life' theory, which divides life into bacteria, archaea (a group of microorganisms that live in extreme environments) and eukarya (cells of which higher life is made), has recently used techniques of gene sequencing with known mutation rates of genes to trace the details of the evolutionary connections that exist between different types of primitive organisms. The situation is analogous to what linguists do in attempting to reconstruct genealogies of ancient languages by mapping the progressive changes that occur in the vocalisation of certain key concepts. In this way Woesse was able to show that the machinery for RNA copying came later rather than earlier in the tree of microbial life, in which case the RNA hypothesis is fatally flawed. Furthermore the same type of analysis, according to Woesse, is beginning to cast doubts on the very concept of a common ancestral cell on the Earth. When different genes are used to map the tree of life a range of equally probable connections arise near its base, and one may not even be able to assert that the crucial genes of eukarya came after archaea in a true evolutionary sequence. There is also talk of a primitive gene pool from which all life may have been derived, and if this pool is thought of as a cosmic pool, the ideas of panspermia, that I shall elaborate further in Chapters 7 and 10, would seem to be vindicated.

It is worth noting that Charles Darwin himself always remained ambivalent on the question as to whether or not there was a first origin of life on the Earth. Attempts to evolve life backwards through an epoch of chemical evolution, for which there is no evidence in the geological record, did not fully meet with Darwin's approval. In a final passage of *On the Origin of Species* Darwin wrote that life was originally 'breathed into a few forms or into one . . .' In 1863 Charles Darwin wrote to Joseph Hooker: 'It is mere rubbish, thinking at present of the origin of life; one might as well think of the origin of matter . . .' It is tempting to link the sentiment expressed here with the assertions of Kelvin and Helmholtz to which we have already referred.

Darwin was not averse, however, to speculate on the possibility of a terrestrial origin for in a later letter to Hooker in 1871 he wrote:

> It is often said that all the conditions for the first production of a living organism are now present, which could ever have been present. But if (and oh! What a big if!) we could conceive in some warm little pond, with all sorts of ammonia and phosphoric salts, light, heat, electricity, &c., present, that a protein compound was chemically formed ready to undergo still more complex changes . . .

It is this statement that seems to have served as an inspiration for all later attempts to reconstruct a primordial soup in a laboratory context. For the reasons given in this chapter, however, all such projects would seem to be inevitably doomed.

In Chapters 7 and 10 we shall show that the ancient idea of panspermia, via the agency of comets, the cosmic dragons, as carriers of life, is one that is destined to become an emergent paradigm of the new millennium. Arrhenius's original defence of panspermia of more than a century ago has been considerably strengthened with many new developments in astronomy, geology and biology, in particular the discoveries of extraordinary survival attributes of bacteria. Arrhenius knew already of the resistance of bacterial spores to conditions of desiccation and their ability to withstand low temperatures. But there was a lot more to follow.

A Planet Grooms for Life

Microbiological research of the past 15 years has shown that many types of bacteria and indeed other microorganisms are remarkably space-hardy, far more so than Arrhenius may have ever imagined. The classes of microorganism known as thermophiles and hypothermofiles thrive at temperatures above boiling point in oceanic thermal vents. Entire ecologies of microorganisms are present in the frozen wastes of the Antarctic ices. A formidable total mass of microbes exists in the depths of the Earth's crust, some eight kilometres below the surface, greater perhaps than the entire biomass at the surface. A species of phototropic sulfur bacterium has been recently recovered from the Black Sea that can perform photosynthesis at exceedingly low light levels, approaching near total darkness. There are bacteria (e.g. *Deinococcus radiodurans*) that thrive in the cores of nuclear reactors and perform the amazing feat of using an enzyme system to repair DNA damage, in cases were it is estimated that the DNA experienced as many as a million breaks in its helical structure.

There is scarcely any set of conditions prevailing on Earth, no matter how extreme, that is incapable of harbouring some form of microbial life. Ultraviolet damage under space conditions is very easily shielded against. A carbonaceous coating of only a few microns thick, which would inevitably arise in many situations, provides essentially a total shielding against ultraviolet light. This shielding occurs in a similar way to the action of sun-tan lotions. Many other tests of the space hardihood of bacterium and viruses have also recently been made. In one such test *Bacillus subtlis* was exposed for nearly six years in space aboard NASA's Long Duration Exposure Facility, and proved to survive. In another experiment bacterial spores recovered from a 250 million-year-old salt crystal from a New Mexico salt pan was revived and cultured. There can be little doubt that microorganisms are ideally suited to be space travellers. They seem designed to survive the rigours of space and may be the closest we can ever come to an example of eternal life.

References

ARRHENIUS, SVANTE (1908). *Worlds in the Making* (trans. H. Borns) London: Harper & Brothers.

DARWIN, CHARLES (1859). *On the Origins of Species*, London: John Murray.

DAVIES, PAUL (1999). Seeding the Universe with Life, *Sky & Telescope*, September 33.

HELMHOLTZ, H. VON (1874) Preface to *Handbuch der Theoretischen Physik* by W. Thomson & P.G. Tait.

MOJZSIS, S.J. *et al.*, (1996). Evidence for life on Earth before 3,800 million years, *Nature*, 384, 55.

MOJZSIS, S.J. *et al.* (2001). A record of Hadean environments in oxygen isotopes, mineral inclusions, trace elements, and the age spectra of ancient (ca 4300 Ma) terrestrial zircons, *Nature*.

SCHOPF, J. WILLIAM (2001). *Cradle of Life: The discovery of the Earth's earliest fossils*, Princeton University Press.

WOESSE, CARL (1998). *Proceedings of the National Academy of Sciences*, USA, 95, 6854.

4. First Steps in a Cosmic Nursery

Regions of lucid matter taking forms,
Brushes of fire, hazy gleams,
Clusters and beds of worlds, and bee-like swarms
Of suns, and starry streams
                    Alfred Tennyson *(1809–1892)*

Theories of origins – the origin of the Universe and the origin of life – have always provoked controversy. Proponents of particular theories have tended to cling to their beliefs with a tenacity akin to religious fervour, often ignoring contrary evidence and dealing with opponents in a ruthless way. Public interest in origins, and more generally in matters that relate to our place in the cosmos, is only natural, and so popular approbation or condemnation may play a part in determining the manner in which theories evolve. Memories of Galileo's trial and his sentence of house arrest in 1633 for maintaining the 'Copernican doctrine' continue to echo down the centuries, as indeed does the more recent Scopes Trial of 1925, in which high school teacher John T. Scopes was tried and fined by the Tennessee Legislature for teaching the theory of evolution.

One often tends to forget that science is an intensely human pursuit and consequently its progress can only be properly understood in relation to the state of society at a particular moment in time. Scientific theories and attitudes are also inextricably linked to our individual as well as collective preconceptions and prejudices, which are in turn a function of the educational system from which we have emerged. My own background and that of my collaborator Fred Hoyle in approaching the problem of the origin of life can be no exception to this general rule. We approached the subject from the standpoint of the physical sciences, in particular from a perspective of astronomy, rather than biology. And it is also the case that we stumbled accidentally on this problem rather than deliberately seeking it out. In Chapters 5 and 6 I show how through seemingly unrelated astronomical studies, we were led logically to this problem and, eventually, to its possible solution as well.

It was in the mid 1950s that radio astronomy began, and amongst its rich harvest were discoveries that would turn out to be germane to the questions that will be addressed in this book. For instance the study of how hydrogen gas is distributed in the galaxy, and the discovery of molecules connected with life, all stemmed from the developments in this field. The mid 1950s also saw the successful completion of work by the two Burbidges, Fowler and Hoyle that led to an understanding of how the chemical elements, including the elements of life, are formed from hydrogen in the deep interiors of stars. At about the same time came the monumental discovery by James Watson and Francis Crick of the double-helix structure of our genetic material – DNA. Shortly afterwards, Frederick Sanger analysed the nature of a protein (insulin to be specific), showing its detailed sequence of constituent amino acids; and Harold Urey and Stanley Miller completed their classic experiments in which they showed how the most basic chemical building blocks of life might be synthesised from inorganic matter.

In astronomy, newly invented techniques, particularly experimental techniques continued to generate an ever-expanding body of factual data. But there was precious little progress in the critical re-appraisal of old hypotheses which was needed now more than ever before. The belief grew that the really important problems were either close to being solved, or that they were really so hard that we should not even begin to worry about them. In practical terms it was a case of turning out more of the same type of factual data, attempting to consolidate existing theories, seeking to tie up the last loose ends. An openness of mind required for a vibrant scientific culture was noticeably lacking.

If the starting point of life involves insuperable odds its origin even once in the entire universe should be regarded as miracle enough. It stands to common sense, therefore, that the origin of life, if there was one, must have involved a physical system of the largest possible scale, a scale that must surely transcend the size of the Earth by an immeasurably large factor. The idea that such an origin must have occurred on a truly minuscule planet such as the Earth borders on the ridiculous. And of course there is no logical requirement whatsoever that life on the Earth must have started on the Earth. The Earth is an open system, and even at the present time it is in receipt of cosmic material at the rate of some hundreds of metric tonnes per day. The entire biosphere of the Earth, which is the sum total of the material of our planet that can support life, is but a figurative drop in the ocean in relation to all the carbonaceous matter that is present in a million galaxies like our own.

Galactic matter that is immediately accessible to biological processing is the material that lies between the stars. An important component of this so-called interstellar matter is in the form of minute solid particles. My collaborations with Hoyle which began in the 1960s started with an attempt to re-examine earlier ideas that related to the nature of this cosmic dust. The firmly held view throughout the 1950s was that the dust was comprised mainly of icy particles similar to the particulate material that existed in the cumulous clouds of the terrestrial atmosphere. Research on inter-stellar dust was thought of at this time as a dull, uninteresting area of science. Most astronomers regarded the presence of dust in the galaxy as a mere nuisance, its only effect being to hinder the observations of distant stars. All that one needed to know was how to correct the measured intensities of starlight for the presence of dust, and for this purpose, simple rules were devised. Nowadays the situation is quite different. Research into various aspects of interstellar dust has become one of the most active areas in modern astronomy. And this transformation began in the mid 1960s starting with the challenges to the ice grain paradigm that Hoyle and I set in train. Before describing our efforts to solve the dust problem, I shall first outline what is known in general terms about the material that lies in between the stars.

We saw in Chapter 1 that the material present in interstellar clouds plays a crucial role in the birth of new stars. Plate 4.1 (top) shows a star field in a part of the Milky Way that lies in the constellation of Sagittarius. Here we see many dark patches and striations running across a background field of stars. We now know that these patches represent dense clouds of obscuring interstellar matter that lie in front of a more or less uniform distribution of stars, blocking out the starlight behind. The dark clouds of interstellar space were spotted by assiduous observers of the heavens for a long time. In 1782 the English astronomer William Herschel was perhaps the first to recognise, and to map these dark clouds systematically, well before the advent of photography.

It became clear through the 1930s that interstellar clouds contained a finely divided particulate or dust component as well as material in the form of a gas. These interstellar dust particles individually have dimensions of a few ten thousandths of a millimetre, just about the size at which light begins to turn corners. Such dust particles turn out to be optimally efficient at extinguishing starlight. Relatively small amounts of dust could produce striking effects in leading to the obscuration of starlight. All this was readily agreed. But over the precise chemical composition of the dust, bitter disputes have raged throughout the twentieth century, and some of these disputes continue even to the present day.

Most of the gas and dust in interstellar space is clumped into clouds of average radius of about ten light years. The typical separation between neighbouring interstellar clouds in the galaxy is about 100 light years. There is, however, a fairly wide dispersion in the sizes of interstellar clouds and in their separations one from another. Some clouds are more compact, whilst others are extended and irregular in structure. The extended clouds appear as giant complexes, exhibiting a great deal of fine-scale structure in the form of cloudlets and filaments, and account for the largest fraction of interstellar matter. These so-called giant molecular clouds are associated with the formation of new stars in the manner described in Chapter 2. A striking example of a cloud complex associated with newborn stars is to be seen in recent Hubble telescope images of the Eagle's Nest Nebula in the constellation of Serpens (Plate 4.1 [bottom]). Newborn stars are embedded in oval-shaped clumps of evaporating interstellar material located at the tips of giant columns of gas and dust resembling elephant's trunks.

An interstellar cloud may contain anywhere from 10 to many millions of atoms per cubic centimetre. But even the higher values in this range of density are considerably lower than the densities attained in the best available laboratory vacuum systems. So it should be remembered that our intuitive ideas of how gases behave under normal conditions in a laboratory could prove wide of the mark when it comes to understanding processes that operate under the rarefied conditions of space.

There is a broad range of temperatures that prevail within interstellar clouds. The denser, darker clouds have gas temperatures as low as five degrees above absolute zero. Interstellar clouds lying in proximity of hot stars could have gas temperatures of tens of thousands of degrees. The temperatures of the dust particles in interstellar space are also variable depending upon their composition, size, and their radiation environment. The dust temperatures which are different from the gas temperatures range from as low as ten degrees above absolute zero to as high as several hundreds of

degrees. We shall see later that the radiation at infrared wave-lengths of these hotter dust grains can tell us important facts that relate to their possible biological origins.

The main constituent by mass of interstellar matter is of course hydrogen. The hydrogen occurs in one of three forms: neutral atomic hydrogen (intact atoms with no electrons lost), ionised hydrogen (atoms stripped of their outer electrons), and molecular hydrogen (atoms paired in molecular form).

The presence of vast quantities of neutral hydrogen was first detected by radio astronomers in the 1950s. An atom of hydrogen in its neutral (uncharged) state comprises a positively charged proton surrounded by a negatively charged electron. Both the electron and proton within the atom possesses a property known as 'spin', which is the analogue in quantum mechanics of the spin of a child's top. Emission of radio waves at a wavelength of 21 centimetres occurs when an electron in a neutral atom flips its spin from a direction parallel to the spin of the proton to a spin in the opposite sense. This radio emission at the wavelength of 21 centimetres has provided an important probe for determining the temperature as well as the distribution in space of neutral atomic hydrogen.

Hydrogen atoms close to hot stars lose their electrons and become 'ionised'. Shells of ionised hydrogen are found in localised regions around hot stars. Within such shells the gas temperatures' could be close to 10,000 degrees.

Molecular hydrogen, H_2 was first detected in interstellar clouds in the 1960s by means of characteristic spectral lines in the

---

Plate 4.1 (opposite page)   *Top*: The Milky Way in the constellation of Sagittarius showing dust clouds from which new stars are born. (Courtesy © Dave Palmer).
Bottom: Dust trailing behind new stars in the Eagle's Nest Nebula (Courtesy © NASA)

ultraviolet region of the spectrum. Such data could only be obtained after the dawn of the Space Age, using observations made from above the atmosphere, because the Earth's atmosphere is opaque to ultraviolet wavelengths. The hydrogen molecule is found to be present in interstellar clouds that are dense enough to screen off the ultraviolet light that would otherwise destroy them. A large fraction of all the hydrogen in the galaxy has been found to be in molecular form, and the total mass of the hydrogen is billions of times the mass of the Sun.

Hydrogen is an important constituent of living material. It occurs in combination with other elements notably oxygen, carbon, nitrogen and phosphorous, and its overall mass fraction in biology is less than 10 per cent. Now let us see what else there is in the interstellar gas clouds besides hydrogen that might be relevant for life. The next most abundant chemical element in the interstellar medium is helium, which accounts for close to a quarter of the mass of interstellar matter. But from our point of view, in relation to life, this element is uninteresting. It exists only in the form of an inert gas that does not react with any other substances, and it can have no relevance for life.

The next most important group of chemical elements is carbon, nitrogen and oxygen that together make up several per cent of the mass of all the interstellar matter. These elements are absolutely crucial for life. Indeed biology depends for its function on the unique range of properties possessed by the carbon atom, that includes its high levels of chemical reactivity and its ability to combine into many millions of interesting carbon-based compounds. Then come the elements magnesium, silicon, iron and aluminium, which again account for one per cent or so of the total mass of interstellar matter. Then a group of elements including calcium, sodium, potassium, phosphorus . . . followed by a host of other less abundant atomic species. We saw in Chapter 1 that all these chemical elements are synthesised from hydrogen in the deep interiors of stars. They are injected into interstellar space through a variety of processes, including mass loss from the surfaces of stars that occurs at certain stages in their evolutionary history. In the case of the more massive stars, the evolutionary end product is a supernova, and it is through supernovae explosions that most of the life-forming chemical elements are expelled into the interstellar clouds.

These elements do not always occur singly in interstellar space. Next to molecular hydrogen the second most abundant and widespread molecule in space is carbon monoxide, CO. A significant fraction of all the interstellar carbon in our own galaxy, as well as in external galaxies, seems to be tied up in the form of this molecule. Next comes the molecule formaldehyde H_2CO, which is present in gaseous form both in clouds of high density as well as in interstellar clouds of relatively low density. In the denser interstellar clouds, particularly in clouds associated with newborn stars, vast amounts of the gaseous molecule water H_2O are found. Water is a crucially important molecule for life, and its close association with newly formed stars and planetary systems could have a direct relevance to extraterrestrial life. It is also of interest that water molecules in dense interstellar clouds often trap radiation at certain microwave wavelengths and give rise to a microwave analogue of the laser phenomenon. Whether this remarkable behaviour is connected to life or not has yet to be fully explored.

The distribution of molecules of various types throughout the Galaxy shows wide variations that depend on the ambient physical conditions such as temperature and density, as well as the proximity of clouds to hot stars. As a rule the denser, cooler clouds contain larger and more complex molecules, whereas lower density clouds and those nearer to hot stars have simpler molecular structures. Many of the complex organic molecules that have been found in interstellar space were discovered using techniques of radio astronomy. A region of the Galaxy that is particularly rich in such molecules is the complex of dust clouds in the constellation of Sagittarius located near the centre of the Galaxy. In this single cloud complex the molecules so far observed include water, carbon monoxide, ammonia, formaldehyde, formic acid, hydrogen cyanide, methyl alcohol, ethyl alcohol, methyl cyanide and most recently of all the amino acid glycine, vinegar and the sugar known as glycolaldehyde. Such molecules, totalling over 80, have been identified by tuning the giant antennae of radio telescopes to precisely the frequencies at which the molecules absorb or emit longwave radiation.

The list of molecules detected so far could represent only the tip of the iceberg. An inherent difficulty of identifications using currently available radioastronomical techniques lies in the require-ment to predetermine the correct radio frequencies that characterise candidate molecules. Once these frequencies are given, the tech-nical problem of looking for the molecules is relatively simple. Molecules that are far more complex than those found so far must almost certainly exist, but their discovery is difficult because of problems associated with calculating the precise radio frequencies that they absorb.

Interstellar organic molecules have been detected by other tech-niques besides radio astronomy, notably using the methods of infra-red astronomy for studying patterns of infrared radiation or heat emission from dust and molecules. We shall discuss such infrared detections more fully in Chapter 5, but note here in passing that an important class of molecule that is found to be present in interstellar space, in quantities that are comparable to carbon monoxide, is the so-called polyaromatic hydrocarbons or PAHs. These molecules are well known byproducts of the combustion of fossil fuels such as occur in automobile exhaust fumes. We shall argue later that even in the Galaxy such molecules are likely to have a biological connotation.

To understand the amount and variety of complex organic molecules that exist in interstellar clouds presents a continuing challenge to astronomers. The orthodox point of view is that complex organic molecules somehow form from individual atoms and simple molecules by means of chemical reactions that take place in the gaseous phase. The density of molecules that follow from any such process represents a balance between formation and destruction, the latter being mostly determined by the ultraviolet radiation to which the molecules are exposed. Because the gas densities in interstellar clouds are so exceedingly low, the reactions occurring between interstellar gas molecules are on the whole far too slow to explain the abundances of complex molecules that are observed. Moreover, in the more tenuous interstellar clouds that are exposed to the full glare of ultraviolet radiation from stars, the dissociation rates are a severely limiting factor in controlling the build-up of organic molecules through gas phase reactions. This is certainly true if the gas atoms and molecules are all electrically neutral: that is to say, they carry no net electric charge. But cosmic rays which penetrate the clouds could ionise or remove electrons from a small fraction of the interstellar atoms. This process could lead to greatly enhanced encounter rates between charged atomic species and small molecules, either uncharged or charged and thus increase the rates at which larger molecules can form. This type of chemistry, known as ion-molecule chemistry, when applied to interstellar conditions has had only a limited measure of success in explaining the presence of certain organic molecules. The more complex organic molecules found in interstellar clouds have to be explained in some other way.

Comparatively high densities of complex organic molecules tend to be associated with regions of the galaxy where new stars are forming at a rapid rate. The Orion nebula is a spectacular example of such a region. It is a stellar nursery where young stars, some with discs of newly formed planetary material around them, can be seen. Large amounts of complex molecules are associated with the denser parts of the Orion cloud complex, and it would be tempting to link the formation of these molecules with the formation of stars, planetary systems, and life. Perhaps all these molecules, including the all pervasive polyaromatic hydrocarbons, are degradation products of biology.

References

HOYLE, F. & WICKRAMASINGHE, N.C. (1977). *Lifecloud.* London: J.M. Dent.

WICKRAMASINGHE, N.C. (1975). *The Cosmic Laboratory*, University College Cardiff Press.

WICKRAMASINGHE, N.C. (1967). *Interstellar Grains*, London: Chapman & Hall.

5. The Cosmic Dust Trail

Ah, make the most of what we yet may spend,
Before we too into the Dust descend.
Edward FitzGerald,
Rubaiyat of Omar Khayyam (1809–1883)

In addition to molecules, interstellar clouds also contain an enigmatic dust component to which I have already referred. One of the great astronomical challenges of the twentieth century was to understand the nature of this cosmic dust, and to determine the circumstances under which it could be formed. Because interstellar clouds turn over into stars with a typical timescale of a few billion years it is necessary to maintain a steady production of dust.

When I began my research in this field with Fred Hoyle in 1960, it was an article of faith amongst astronomers that these dust grains were comprised of dirty icy material – frozen water with perhaps a sprinkling of other ices – ammonia-ice, methane-ice and a trace quantity of metals. It was further maintained that such icy grains must condense, more or less continuously from the gaseous atoms of carbon, oxygen, nitrogen and other metals supplied from mass loss in stars and supernovae, and from the molecules that already existed in interstellar clouds. In 1960 the ground rules for this area of astronomy appeared to have been laid down in two classic papers and a thesis written in the mid 1940s by the Dutch astronomer H.C. van de Hulst. (Incidentally, it was van de Hulst who predicted and first observed the 21 centimetre line of neutral hydrogen in the Galaxy, a subject to which we referred in Chapter 4). Impressive though van de Hulst's contributions to radio astronomy had been, we soon discovered that his ideas relating to the nature of cosmic dust were flawed in several important respects.

After I had completed a detailed review and reappraisal of the ice grain theory I went to Leiden in 1963 to discuss these matters with van de Hulst. It took me only a few minutes to realise that the great man had lost all interest in this subject, and so he promptly directed me to his colleague, J. Mayo Greenberg, in New York, who had taken up the cudgel of defending the ice grain theory against all its critics.

Our first major objection to the ice-grain model became apparent as soon as we began to think seriously about it. Even though the atomic constituents of water molecules, hydrogen and oxygen, exist in abundance in interstellar space, the first requirement was for these atoms to combine in the gas phase to form H_2O. That process is by no means straightforward at the very low ambient densities that prevail in interstellar clouds. Water molecules, wherever they already exist in interstellar space, could of course condense as solid ice on pre-existing dust grains. But if there is no available supply of such pre-existing dust grains or grain nuclei, how could condensation occur? The requirement for any ice condensation process in interstellar space is that condensation nuclei be injected into the interstellar clouds at a steady rate. Again, in view of the very low gas densities that are present, these nuclei could not be generated sufficiently fast from within the interstellar medium itself. The problem is similar to that of the seeding of water-vapour clouds in the Earth's atmosphere. One could have highly supersaturated clouds in the atmosphere, but no rain will fall unless condensation nuclei are somehow supplied.

Theories of cosmic dust formation are all well and good, but how do such abstract theories account for the astronomical facts? It is exceedingly difficult, if not impossible, to investigate in the laboratory any theory of how dust may form under the tenuous conditions of space and over astronomical timescales of millions of years. But if we have arguments of plausibility to support a particular model of cosmic dust, for example the ice grain model, we could proceed to examine the consequences and predictions of such a model in terms of astronomical observations that could be made. This has been my own personal approach to this subject, and one that has occupied a good fraction of my career in astronomical research. If we think that the dust in the dark clouds of space is made of a particular substance – e.g. ice, or rock or soot – it is possible to measure the properties of this material in the laboratory, and then use well attested theories of light transmission to predict what this dust would do to starlight as it passes through interstellar space. Then we can turn to the telescope to check a prediction based on a particular dust model.

A source of natural light, including starlight, is a mixture of colours (wavelengths) ranging from ultraviolet, violet and blue through into the red and infrared. The weighting of the various colours of the spectrum for any source of natural light towards either the red or the blue depends upon the temperature of the source: the cooler sources are redder, the hotter sources are bluer. Hot blue stars appear redder and *apparently* cooler when they are viewed through clouds of interstellar dust. The reddening occurs by a process similar to the reddening of a street lamp seen through a fog, or the reddening of sunlight in the evening sky. Submicrometre-sized dust, that is to say, particles with diameters of about a tenth of the thickness of a human hair, scatter and absorb the blue component of light from any source more strongly than the red component. So a source of natural light would be apparently reddened. This would happen irrespective of what the dust is made of, provided their particle sizes were in the submicrometre range.

It is also a known fact that for fixed masses of the dust, those particles which have circumferences close to the wavelength of the light have a stronger scattering effect than either particles that are a lot larger or a lot smaller. It is as though a solid particle tunes up with light whose wavelength is comparable to its own girth. Such an effect could be demonstrated to an audience in a powerful way with a slide projector and a source of smoke (e.g. a cigarette). A puff of smoke that is made to intercept the light beam of the projector will dramatically darken the screen. This shows that particles of smoke have sizes that are on average comparable to the wavelength of the light shining on them.

The first attempts by astronomers to obtain quantitative measurements of the dimming or extinction of starlight as it traverses through clouds of interstellar dust were made in the mid 1930s. Using a number of different techniques, some of which were quite ingenious for their time, it was shown that for starlight of a particular colour, at a typical blue wavelength of say, 4,500 Angstroms, the dimming amounted on the average to a halving of its intensity for about every 3,000 light-years of traversal through interstellar space. Another important property of the dust discovered about the same time related to the reflectivity (albedo) or shininess of individual dust particles. Studies of the way in which dust clouds reflected starlight, when they happened to lie close to stars, were interpreted in a straightforward way to show that interstellar dust must be made of material that had rather poor absorption properties at visual wavelengths. This data already restricted permissible models: for instance they could be comprised of ice, wax, organics or minerals, but not metals.

From the astronomical data that was available in the 1940s on the dimming of starlight, it would seem remarkable that reliable estimates could be made of the total amount of dust that existed in interstellar space in the form of submicrometre-sized particles. The result is simply this: Throughout the entire volume of the disc of the galaxy, where most of the stars reside, the mass of dust is about ten million times the mass of the Sun. This estimate turned out to be relatively insensitive to the issue of what the particles are actually made of, so long as they comprised mostly non-absorbing, nonmetallic type materials.

How does this calculated mass of dust relate to the total mass of interstellar matter in the galaxy? By combining estimates of the total mass of interstellar material in the Galaxy with the known proportions of the different chemical elements, two further conclusions could be reached. The dust in the Galaxy accounts for one percent or so of all interstellar matter, and what is more, we require that a very large fraction, as much as a quarter, of all the carbon, nitrogen and oxygen in interstellar space is tied up in the form of cosmic dust. These simple and incontrovertible facts turn out to be of crucial importance in arguments that we shall now develop.

Broadly speaking there are two different classes of chemical composition that are possible for the bulk of the material of interstellar dust. Combinations of carbon, nitrogen and oxygen with hydrogen could occur as frozen inorganic materials, producing a model generally similar to that originally proposed by H.C. van de Hulst. Alternatively, one could envisage dust grains of a predominantly carbonaceous composition, including also the possibility that some of the carbon is in the form of graphite particles, and some in the form of the stable C_{60} cages known as buckyballs, and a fraction in the form of nanometre-sized diamond particles, which is yet another form of carbon.

Throughout the 1950s, 1960s and 1970s astronomers were able to make increasingly accurate determinations of the extinction or dimming of starlight by interstellar dust, not just at one wavelength but over a wide range of wavelengths. The wavelength base for these studies was gradually enlarged from just three visual colours in the early 1950s to a continuum of wavelengths that ultimately spanned the spectrum from the infrared into the far ultraviolet. The constraints on allowable models of the grains consequently became more and more rigorous with the passage of time. The earliest astronomical studies suggested that the dimming of starlight measured on a logarithmic scale was roughly in inverse proportion to the wavelength of light. At the time Hoyle and I turned to the problem of interstellar dust, the range of wavelengths covered by the observations was still rather limited, extending from about 10,000 to about 3,000 Angstroms, that is to say, from the near infrared to the very near ultraviolet. Over this range the extinction curve of starlight departed from an inverse proportional dependence on wavelength as one moved away from visual wavelengths towards the infrared as well as towards the ultraviolet. But even so it was not possible to use the extinction curve as a diagnostic of any particular grain material, as long as it fell within the permissible categories that we have discussed.

A point that impressed us forcibly at the outset was that for any given dust grain composition the extinction curve, or the colour dependence of the attenuation of starlight, as it was known in 1962, demanded an exceedingly rigorous definition of particle sizes. Because the law of extinction was found to be the same in all directions in the sky, the requirement was that a size parameter for the dust had to be finely tuned everywhere in the galaxy. For instance, for the van de Hulst model of ice grains an average size parameter (about 3,100 Angstroms or three tenths of a micrometre in radius) had to be specified and fixed to a precision level of better than three per cent. We found such a restriction of size very hard to accept. If icy particles did indeed condense in interstellar clouds, it would be implausible to require their sizes to be so rigidly fixed on a galaxy-wide scale. There would surely be a wide range of ambient conditions that prevail within interstellar clouds, some clouds being denser than others, and the lifetimes of clouds would also vary by a large factor. Any interstellar condensation process that operates must clearly lead to a wide range of resulting dust grain sizes from one galactic location to the next. This difficulty has been glossed over by successive generations of astronomers who remained wedded to the concept of icy grains in space.

Our tentative steps towards the cosmic dragons had an unlikely beginning: in 1962 Hoyle and I challenged the dirty ice-grain model in a joint paper published in the *Monthly Notices of the Royal Astronomical Society*. In it we advocated a shift away from a 20-year strong tradition of volatile icy grains in space to sturdier carbonaceous grains. In the original carbonaceous model that we considered we opted for the most highly reduced form of carbon – graphite.

Because of the difficulties associated with dust grain formation in interstellar space we decided to transfer the venue of grain condensation to the envelopes and atmospheres of cool giant stars – stars with distended atmospheres, and relatively low surface temperatures. A class of giant star, the carbon stars, possess surface temperatures that vary in the range, 1,800–2,500 degrees Kelvin, the variation taking place over a cycle that lasts for about a year. These stars also possess an excess of carbon over oxygen in their atmospheres. We were able to show that during the low temperature phase of the stellar cycle carbon must condense into small particles, the earlier nucleation problems of interstellar space being overcome here because of the very much higher densities that prevail. Furthermore, we showed that the condensed dust particles could be driven away from the parent star because of the pressure exerted by starlight shining behind them.

There was substantial evidence to suggest that a process like this actually takes place in certain stars. The carbon star R Corona Borealis provided spectacular evidence of carbon dust formation: the star is seen to be smoking like a chimney. There were also similar processes taking place in another class of cool giant star – the so-called Mira variables. Here the abundance of oxygen in the atmosphere exceeds that of carbon, and the resulting dust particles are mineral silicates rather than carbon. While there was no doubt that a fraction of the interstellar dust may have a stellar origin in this way, the big question that remained to be answered was how much? It seems doubtful now that processes of this kind could account for more than a small fraction of the grand total of 10 million solar masses of dust that is actually found to exist in interstellar space.

Our first public confrontation with the proponents of the ice-grain theory took place at an international conference held in 1965 in Troy, New York. I made a presentation of the arguments against ice grains and in favour of carbon grains, while Mayo Greenberg passionately defended the ice-grain theory. A paper by Greenberg argued that the best fit by far for the entire range of observational data was obtained using ice cylinders 'of infinite length, with a tri-modal distribution of radii': a situation reminiscent of Ptolemy's epicycles, one might say. Greenberg's position in relation to water-ice grains became manifestly untenable at this conference, after T.P. Stecher of NASA Goddard Space Flight Centre had shown that the dust possessed an absorption feature in the ultraviolet near the wavelength of 2,200 Angstroms. Such a feature did not fit the ice grain model.

The pioneering work of T.P. Stecher was taken many steps further a year or so later when more accurate ultraviolet observations became available using satellite-borne telescopes such as the IUE (International Ultraviolet Explorer). New observations obtained by R.C. Bless and B.D. Savage and later by K. Nandy for a very large number of stars confirmed Stecher's original results. In addition to a broad absorption band of the dust centred at 2,175 Angstroms, they found a continued rise of the average extinction curve into the further ultraviolet.

It was a remarkable coincidence that this 2,175A feature matched precisely the absorption behaviour of graphite particles with radii of about 200 Angstroms. I had predicted the existence of such a feature for graphite ahead of the astronomical observations becoming available. When the observations came along it seemed natural to congratulate ourselves and to attribute this interstellar feature to graphite dust. There was general acceptance of this assignment and most astronomers still believe that our original graphite identification is valid. Hoyle and I now think differently, however. Calculations show that the graphite particle extinction feature can be centred on 2,175 Angstroms, as the astronomical data requires, only if the particles are perfectly round with radii very close to 200 Angstroms. A problem immediately arises here because real graphite does not form as round spheres – rather it forms as flakes or needles, and for such particle shapes no agreement with the data is possible. We shall argue later that the observed astronomical feature at the wavelength of 2,175 Angstroms is more likely to be due to molecular absorptions in complex organic molecules. Such organic molecules, called aromatic because they are normally endowed with aroma, contain several joined-up hexagonal carbon rings within which ultraviolet absorptions occur.

Another presentation of historic importance presented at the New York conference of 1965 was by a chemist, Fred M. Johnson. Johnson argued that over 20 diffuse interstellar absorption features, the strongest of which is centred at the blue wavelength of 4,430 Angstroms, and which thus far have remained unidentified, might be due to a highly complex aromatic molecule associated with the cosmic dust. This molecule, he argued, was most likely to be a porphyrin similar to chlorophyll, the green colouring substance of plants. Chlorophyll is of course a vitally important component of terrestrial biology the molecule responsible for photosynthesis, the process that lies at the very basis of our entire ecosystem on the Earth. It is my view that Johnson's explanation remains to this day a promising solution to the still unsolved question of the diffuse interstellar bands. At the time, however, Johnson's paper was widely ridiculed, it being argued that organic molecules of any sort will not survive in interstellar space. Johnson's was an idea well ahead of its time, although now in AD2001 organic interstellar molecules are taken for granted, and we can see that he may well have been a prophet who ushered in the dragons to their dens.

My initial objections to the dirty ice model of cosmic dust which were mainly based on the difficulties associated with their formation, were strengthened by developments that took place in 1965. At about this time the first observations became available of the spectra of stars over an infrared waveband near the wavelength of three micrometres. This is precisely the part of the spectrum where water molecules in ice crystals show a strong resonant absorption, and if ice is present within interstellar dust clouds, an absorption centred at a wavelength of 3.1 micrometres must be observed. Even a small fraction of ice, if it is present in the interstellar dust, should show up unmistakably at this wavelength.

R.E. Danielson, N.J. Woolf and J.E. Gaustad were the first astronomers to search for this water-ice absorption band in the infrared spectra of stars that had been substantially dimmed by interstellar dust. The feature was not found. Its absence could be shown to imply that water-ice particles, if they exist at all, must make only an extremely minute contribution to the dust of our Galaxy. Although it is true that water-ice absorption bands have subsequently been detected in a number of astronomical sources, these are found to occur only in relatively localised regions, in dense clouds associated with newly formed stars, and in proto-stars – dust clouds that are on the verge of collapse towards forming new star systems. In such situations it would be reasonable to infer that water-ice condensation would occur as mantles or coatings on already existing dust particles, thereby growing in size. No substantial water-ice absorption has been detected in the general tenuous interstellar medium.

There was another problem with the graphite model as far as the ultraviolet extinction is concerned. Although graphite in the form of 200 Angstrom radius spheres fits the interstellar absorption band near 2,175 Angstroms, its extinction efficiency falls off sharply at much shorter ultraviolet wavelengths. To explain the continued rise in the far ultraviolet extinction, a separate dust component in the form of very small (500 Angstrom-sized) non-absorbing particles seems to be required.

The position we had reached now, in the mid 1960s was that although ice grains were decidedly ruled out, a simple graphite model of cosmic dust also faced serious difficulties. With no one type of cosmic dust particle capable of explaining the whole range of available data, we turned next to exploring the properties of combinations of materials. We considered particles in the form of graphite cores surrounded with icy outer mantles (core-mantle particles), and mixtures of several distinct particles species. In 1968 our favourite grain model involved a mixture of particles of graphite, silica and iron.

It was at about this time that a new observational development intervened. Thus far the infrared observations of stars had led only to negative results – there was no infrared spectroscopic feature of ice, therefore no ice. With further advances in the techniques of infrared spectroscopy it soon became possible to obtain high-resolution spectra of astronomical sources over a much wider range of wavelengths. This included high-resolution data over the infrared waveband ranging from about 8 to 13 micrometres. In 1969 a group of astronomers based at the University of Minnesota, led by E.P. Ney, made the first positive detection of an infrared spectroscopic feature in the interstellar dust. They discovered a broad spectral feature over the wavelength range of 8–13 micrometres, centred on the wavelength of 10 micrometres. This feature was clearly a signature of interstellar dust. It was seen in emission in a variety of astronomical sources, including cool oxygen-rich giant stars and in the dust within the Trapezium region of the Orion nebula. At first sight it appeared that this feature might signify the presence of dust in the form of silicate minerals, because silicates do indeed possess absorption bands close to the 10 micrometre wavelength. On closer examination we discovered that no known silicate can reproduce the important details of the observed astronomical feature to a sufficient degree of accuracy.

By the end of the 1960s interstellar dust grains were assuming an importance in astronomy that could not have been guessed at the start of the decade. This explosion of interest in the dust led to its inclusion in discussions over a wide range of problems in astrophysics. It was becoming clear that dust played a crucial role in controlling many phenomena – the formation of hydrogen molecules, the condensation of gas clouds to form stars, as well as in the process of degrading visual and ultraviolet starlight to produce infrared radiation and perhaps radiation at even longer wavelengths. Many external galaxies were discovered to be strong emitters of infrared radiation, with some galaxies, such as the Seyferts, being far more luminous at infrared wavelengths than in visual light. Such observations proved that cosmic dust was not confined to our Galaxy alone. Dust with properties very similar to the properties of galactic dust exists outside our Galaxy, and on a truly vast scale. Many extragalactic objects, including the so-called starburst galaxies, which as the name implies are hugely active in the formation of new stars, were found to have dust grains bearing the same spectral signatures as galactic dust. This similarity extended over the 8–12 micrometre waveband as well as at other infrared wavelengths and in the visual and ultraviolet.

The astronomical source of 8–12 micrometre radiation that is the most straightforward to interpret and model is the Trapezium region of the Orion Nebula in our own galaxy. Here we find a 'thin' extended cloud of dust heated to temperatures in the range 150–300 degrees above absolute zero. The infrared radiation emitted from this dust shows a broad peak centred at a wavelength close to 9.7 micrometres. The observations are shown as the points in Fig. 5.1. The two curves show the optimal fits possible for silicate models, including both hydrated and amorphous silicate, both of which are clearly seen to be grossly inadequate. Many other silicates have been considered to date, including moon rock and a variety of minerals, some which have been irradiated in the laboratory with a view to simulating space conditions. But the mismatch depicted in

Fig. 5.1 remains essentially the same and endemic to all silicate models of the dust.

Between the wavelengths of 8 and 9 micrometres silicate minerals do not appear to have an adequate strength of absorption to account for the astronomical data. To overcome this difficulty a somewhat dubious practice has come into vogue. For reasons that are not at all clear, astronomers have come to think that they are bound to identify silicates from the data of Fig. 5.1. So 'silicates' and

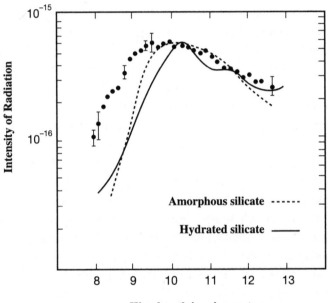

Wavelength in micrometres

Figure 5.1   The mismatch between silicate models of cosmic dust and the emission at infrared wavelengths from dust in the Trapezium Nebula. This mismatch that was endemic to all silicate models of dust pointed the author in the direction of organic and biological dust. A model involving a mixture of diatoms (siliceous algae) and bacteria leads to a curve that passes precisely through all the data points.

'Trapezium material' have been deemed to be interchangeable, as if by some sort of divine decree. To many astronomers an 'astronomical silicate' has become synonymous with the material actually present in the Trapezium nebula. The plain truth is that a silicate is a silicate is a silicate, with properties that are wholly determinable in the laboratory. If after 30 years of experimentation in the laboratory no matching silicate has been discovered, it is high time that an alternative nonsilicate model is considered.

Throughout the period 1972–1974 I continued to be haunted by the glaring mismatch between silicate models of the dust and the astronomical data as seen in Fig. 5.1. Early in 1974 I decided to explore a dramatically different avenue for grain composition. With the rapidly expanding list of organic molecules in space it seemed natural to wonder whether much of the dust could be organic as well.

In view of the ubiquitous presence of the molecule formaldehyde – which occurs both in diffuse interstellar clouds as well as in dense clouds – this molecule presented itself as an obvious first candidate to be considered. Could the interstellar formaldehyde gas be connected in some way with a solid particle of the same composition? I decided to examine the conditions under which formaldehyde gas in interstellar clouds might form into a stable polymer, a chain of strongly linked formaldehyde molecules, and to investigate how this material might contribute to absorptions and emissions at infrared wavelengths. These thoughts began to take shape throughout the summer of 1974 when I was visiting the University of Western Ontario in Canada, and several weeks were taken up in a search for relevant data at the library there. I discovered quickly that the formation of formaldehyde polymers could indeed occur under conditions that prevail in dense interstellar clouds. Furthermore polyformaldehyde grains were found to be better suited than silicates to account for the 8–12 micrometre interstellar emissions, not only in our own galaxy but in external galaxies as well. These conclusions were published in the *Nature* (6 December 1974), this paper becoming the first ever publication to propose that interstellar grains were comprised of complex organic polymers.

For the next two years, 1974–1976, I became increasingly obsessed by the idea of organic polymers being processed on a huge scale in the galaxy. Models involving co-polymers – mixed chains of formaldehyde with other molecules – as well as polymer mixtures resembling tars, were all explored. During this period Hoyle and I were temporarily out of contact. he had been spending considerable periods of time at the California Institute of Technology following his departure from Cambridge.

Early in 1976 my work on interstellar polymers began to veer dangerously towards a possible biological connotation. The program of research that I embarked upon in 1960 was leading towards one of the most important problems in science: the origin of life. To seek moral support from my mentor as much as for technical guidance, I initiated a transatlantic correspondence with Hoyle. Hoyle received a series of letters from me describing the trend of the polymer work, and including suggestions how this may imply the origin of life on a colossal extraterrestrial scale. It was perhaps fortunate that Hoyle himself had an instinctive dislike of the standard Haldane–Oparin model of the origin of life on the Earth. In his *Frontiers of Astronomy* published in 1954 he had suggested for the first time that the Haldane–Oparin–Urey–Miller picture would have a far greater plausibility if it were extended to include the entire solar nebula, the cloud of material from which the Sun and planets formed. So the proposition I had placed before him in 1976 did not come as a complete surprise. But even so his immediate reaction was to call for caution, demanding more evidence before becoming convinced. At first Hoyle played the important role of devil's advocate in our dialogue, but his interest in this problem was sparked, his convictions grew in strength and our collaborations resumed with full steam. On 4 November 1976 the idea that a prebiotic chemistry involving an evolution within clumps of tarry interstellar grains was published in the form of a joint letter to *Nature*.

It was at about this time that we began examining critically the strengths and weaknesses of the reigning paradigm relating to life's origins. This paradigm essentially asserts without proof that

> in the beginning a primordial soup developed in the Earth's oceans, and life began as a result of chemical transformations of molecules in the soup . . .

Some form of this statement, deriving its inspiration from Darwin's letter to Hooker of 1871, referring to 'some warm little pond', has been repeated in the first chapter of every textbook in biology since the 1940s. On closer inspection it soon becomes clear (as was seen in Chapter 3) that the standard theory on life's origins had very little to commend it. In contrast, an origin of life that involved a cosmic system had a lot going in its favour. With cosmic biology at the back of our minds we began to re-examine critically a wide range of astronomical data. In particular we looked for evidence relating to the existence of biochemical substances as opposed to simple organic or prebiotic polymers. Recalling that the fit of the polyformaldehyde-Trapezium spectrum published in 1974 was by no means perfect, we spent many weeks early in 1977 searching the spectroscopic literature for other plausible organic polymers and organo-siliceous compounds. At last we chanced upon an infrared spectrum of cellulose – a polysaccharide (an arrangement of sugar molecules into ring structures), the substance that gives strength to the stalks of plants, and the most common of all biological polymers on Earth. A glance at the cellulose spectrum showed it to have properties of greater interest in the infrared than anything we had seen before. At long wavelengths the absorption due to carbon-oxygen-carbon bonds in the 8–10 micrometre region matched the data from the Trapezium nebula, while at short infrared wave-lengths there was a broad absorption near three micrometres that also matched the spectra of certain dense clouds of interstellar dust.

The uncannily close agreements that emerged between our calculations for polysaccharides and the infrared spectra of certain types of astronomical sources made it amply clear that cosmic dust grains must have a biochemical component, or at least a component that was indistinguishable from a biochemical one it its spectral properties. The infrared sources that best matched the properties of polysaccharides were mostly associated with localised dense clouds, usually connected with proto-stars or new-born stars. Throughout much of 1977 and in the early months of 1978 we toyed with the idea that polysaccharide-like grain material might form through nonbiological means in material that flowed out from massive young stars. We pursued such ideas vigorously, and our calculations were published in *Nature* and in the *Monthly Notices of the Royal Astronomical Society*, just as we had done at earlier times. As long as we kept well clear of extraterrestrial biology as such, and only discussed pre-biology, we had no serious trouble getting our work published in this way. And up to this moment our work, although not in the main stream of astronomy, continued to receive the approbation of the astronomical community. The approach to cosmic biology proceeded inexorably, however, and we soon began to alienate friends and colleagues. Their vision regrettably turned out to be more limited than ours, and ultimately they were proved to be wrong.

References

HOYLE, F. & WICKRAMASINGHE, N.C. (1991). *The Theory of Cosmic Grains*, Dordrecht: Kluwer Academic Publishers.
WICKRAMASINGHE, N.C. (1967). *Interstellar Grains*. London: Chapman & Hall.

6. A Galaxyful of Bacteria

How rough a sea –
And, stretching over Sado Isle,
The galaxy
          Matsuo Basho *(1644–94)*

The origin of life cannot be understood separately from the origin and structure of the universe as a whole. If the origin of life preceded the origin of the Earth and of the solar system, as we have indicated in an earlier chapter, how far back in time did the ultimate event of origination occur? This question logically brings us into contact with cosmology, a subject that we briefly touched upon in Chapter 1.

The standard cosmological theory of the origin of the Universe is the so-called Big-Bang theory where the entire Universe came into being from nothing some 13–15 billion years ago. It is a matter historical curiosity that this theory's – name Big Bang – was origianlly one of derision. Fred Hoyle, who provided this label, together with Thomas Gold and Hermann Bondi, had proposed an alternative Steady-State theory of the Universe in 1948, in which the Universe expands, but as it does so, new matter is created at a steady rate, so as to maintan an average constant density. The mean creation rate of matter required in the original Steady-State theory was estimated as being equivalent to the production of one new hydrogen atom in a typical-sized assembly hall every 1,000 years –

a very slow rate indeed. Modern versions of the theory consider an endless sequence of oscillations of the Universe with creation of new matter taking place, not uniformly everywhere and at all times, but only during the densest epochs, in objects such as black holes. These ideas have been discussed during the past years by Hoyle, Burbidge and Narlikar in a theory referred to as the Quasi-Steady-State Cosmology. It is this later revised version of the Steady-State theory that appears to be in accord with all the modern astronomical facts.

Steady-State cosmologies in general share the property of an open timescale and an infinite matter content that make them better suited to deal with the problem of the origin of life. The super-astronomically improbable event of the first origination of life would have a chance to occur sometime, somewhere in such an infinite framework. The idea of a Universe with an open timescale remains, however, as unpopular now as it was in the sixteenth century. The Italian philosopher Giordano Bruno (1548–1600) has earlier rejected the traditional concept of a closed finite Universe with nothing existing beyond the 'sphere of the fixed stars'. In its place he had proposed a theory of an open infinite Universe whose broad reaches were characterised by evenness. For this heresy he was burned at the stake in 1600. In the year 2001 we have come a little way from such extremes of conduct, but attitudes to cosmology still tend to be characterised by intransigence and intolerance. The persecution of heretics is subtler nowadays.

Nobody takes lightly the prospect of walking into exile, albeit scientific exile, and one that is self-imposed. Yet by the autumn of 1977 it appeared to Hoyle and myself that we had no option. If we were to pursue the truth with unrelenting determination, we simply went where the trail led.

Whilst we continued to work out details of the properties of organic polymers in interstellar space as well as in comets, it became increasingly difficult for us to avoid the conclusion that the Haldane–Oparin primordial soup theory was seriously flawed, and that life must have arrived at the Earth from space. Of this we became convinced by the end of 1977. Such ideas ran so much against the grain of conventional thinking at the time that we faced serious difficulties in having our work published through customary channels. Some referees of the journal *Nature* went so far as to state that there was no firm evidence for organic grains in space, and others even claimed that complex organic molecules could not survive in space. These assertions were of course a blatant travesty of the truth.

At this point two far-sighted individuals came to our rescue. C.W.L. Bevan, then Principal of University College, Cardiff, chose to support our endeavours wholeheartedly by publishing our ideas in the university press, and Zdenek Kopal, founding editor and editor-in-chief *Astrophysics and Space Science*, and Professor of Astronomy at Manchester University offered to publish our ideas in his journal.

Our first unequivocal pronouncement on the extraterrestrial nature of life appeared in a special Cardiff University Press Preprint in October 1977 and was repeated in *New Scientist* in November 1977. In it we listed our objections to the standard Haldane–Oparin model, and concluded:

Ultraviolet sunlight incident on melted cometary ices could induce polymerisation reactions leading to the elaboration of prebiotic molecules into more complex structures such as poly-nucleotides, polypeptides and porphyrins. And such macro-molecules which are produced in one cycle could hold 'survival patterns' in a frozen state until the next perihelion passage when simpler molecules could be recast into 'predetermined' struc-tures. The first living organisms may have evolved in response to these selective pressures. Photosynthetic bacteria, which could oxidize hydrogen sulphide anaerobically, may have been the result of this evolution. Such bacteria are known to be viable in situations where the sunlight intensity is as low as 1/10 of that required for the survival of normal photosynthetic bacteria. This property would be an asset for primitive cometary organisms which must attempt to maximise their 'effective life-span' by operating under conditions of the lowest possible sunlight intensity compatible with immersion in liquid water. The requirement for survival at perihelion, when the temperature is high, favours the evolution of thermophyllic bacteria – similar to those found in certain terrestrial hot springs . . .

Our initial proposal of starting life *de novo* in individual comets did not have too long a persistence in our thoughts, for reasons given later. However, ideas similar to those contained in the above quotation dating back to 1977 have become more or less a consensus view at the present time.

In the same publication we also began to develop a second strand to our arguments relating to extraterrestrial life.

A direct consequence of our theory is that attempts at starting life anew from prebiotic molecules must continually be occurring on the surfaces of comets, particularly on 'new' comets which have long periods of revolution around the Sun. At regular intervals the Earth must pick up debris from such comets in the form of micrometeorite showers: and at more regular and frequent intervals it picks up materials from shorter-period comets. The question arises whether encounters with such material, containing fresh attempts at starting life, could have a deleterious effect on indigenous terrestrial biology. The likely answer to this question is in the affirmative . . .

We contend that extraterrestrial encounters of this type lead to the injection of disease-causing bacteria and viruses on to the Earth.

These ideas were still in embryonic form at this stage, but already in 1977, we thought of a possibility of testing the extraterrestrial life theory by looking at the patterns of viral and bacterial diseases on the Earth (see Chapter 10).

The sequence of ideas relating to theories of interstellar dust we had developed defined a route that soon became inevitable. From the outset we realised that carbonaceous dust in the galaxy had to involve more than a third of all the carbon atoms in interstellar space. For the composition of this dust we progressed from a carbonaceous model in the form of graphite, to organic polymers, and thence to complex biopolymers. These organic polymeric particles, that were now in evidence everywhere in the galaxy, had to possess an average radius of about a third of a micrometre, about the size of a typical terrestrial bacterium.

Our first thoughts were that these organic particles might be the result of some sort of prebiotic process, a process perhaps similar to that described in our earlier quotation, and one which might lead eventually to the origin of life in comets. But a nagging question remained. If the interstellar organic polymers were all the result of prebiology, how could such a process be so amazingly efficient in converting so large a fraction of the available carbon into organic particles of bacterial size? After many months of fumbling through a sequence of models, all of which were proving to be woefully inadequate, an astounding thought occurred to us in the middle of January 1979. In the gigantic clouds of interstellar dust could we be witnessing no less than the operation of biology itself? Could the interstellar medium be choc-à-bloc, not simply with the building blocks of life, but with the end products of the living process as well? And this would then be required to happen on an un-imaginably vast cosmic scale. Interstellar grains must surely be bacteria – albeit freeze-dried, perhaps mostly dead.

Next, the bacterial model of interstellar dust had to be tested as matter of urgency. There were calculations as well as laboratory experiments that needed to be done. How do bacterial grains fare in relation to the observed extinction or dimming of starlight, particularly over the visual spectral region in which the extinction curve is so startlingly invariant? The first things we needed to check were the sizes of bacteria and their optical properties. It took us only a short while to discover that the sizes of spore-forming bacteria on the Earth were just right to be relevant in a cosmic context. Next we were able to deduce that the bacteria, when they are fully dried out, as they would be in interstellar space, would have cavities that amounted to some 60 per cent of the total volume – the space that the water in them had occupied. It was a merit of our model that there were no other parameters that remained free to adjust. This situation was in sharp contrast to all the inorganic models that were considered earlier, where there was always at least one arbitrary parameter to be fixed. With the benefit of long years of experience of doing similar calculations, we were able to check the extinction behaviour of our new model within a matter of hours. The result along with the relevant astronomical data obtained by Kashi Nandy at visual wavelengths is shown in Fig. 6.1. This exceedingly close agreement between the observations and our bacterial model convinced us that we were on the right track. And then there was no turning back.

We distributed our results to our colleagues at the beginning of April 1979, and wrote a technical paper 'On the nature of interstellar grains' for *Astrophysics and Space Science*. It now became more urgent to discover what further checks needed to be made. Predictions of the behaviour of a bacterial model at infrared wavelengths had to be done, and these might then be looked for in astronomy. But this required both experimental work in the laboratory and work at a telescope as well.

Shirwan Al-Mufti, research student at Cardiff, was given bench space and laboratory facilities in the Biochemistry Department there, in order to undertake spectroscopic studies of biological samples. Experiments began to yield important results in the first few months of 1981.

These experiments involved desiccating bacteria, such as the common organism *E. Coli*, in an oven in the absence of air, and

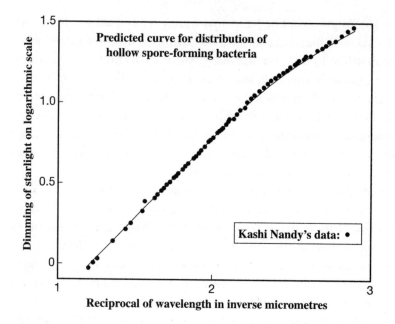

Figure 6.1   The fit of the astronomical data on the dimming of starlight by interstellar dust (points) to a bacterial model (curve). This data set could not be matched to anything like the degree of precision shown here by any of the competing inorganic or nonbiological models for cosmic dust.

measuring as accurately as possible the manner in which light at infrared wavelengths was absorbed. The normal technique for doing such a measurement involved embedding the bacteria in discs of compressed potassium bromide and shining a beam of infrared light through them. The standard techniques had to be adapted only slightly. There was the need to match the interstellar environment which involved desiccation, and the spectrometer that was used had to be calibrated more carefully than a chemist would normally do. When all this was done it turned out that a highly specific absorption pattern emerged over the 3.3–3.6 micrometre wavelength region, and this pattern was found to be independent of the type of microorganism that was looked at. Thus whether we looked at *E. Coli* or dried yeast cells it did not matter. This invariance came as a welcome surprise.

Our newly discovered invariant spectral signature was a property of the detailed way in which carbon and hydrogen linkages were distributed in biological systems. So it followed that if interstellar dust was comprised of bacteria, this was clearly the spectroscopic signature that we should look for.

If astronomy turned up with a totally different profile, then our model would have been falsified. Because we knew the exact amount of bacteria in our laboratory sample causing a measured amount of absorption, we could also determine how strong the absorption by bacterial dust should be at any particular wavelength. From our measurements it turned out that this absorption band was itself intrinsically weak. Thus the longest possible path length through the galaxy was needed to have any hope of detecting the effect. The longest feeasible path length through interstellar dust that existed within our own galaxy was defined by the distance from the Earth to the centre of the Galaxy. Was there any source of infrared radiation near the galactic centre that could serve as a beacon for detecting interstellar bacteria?

My brother D.T. (Dayal) Wickramasinghe, Professor of Mathematics at the Australian National University in Canberra, was also an astronomer and frequently used the 3.9 metre Anglo-Australian Telescope for a variety of astronomical projects. The AAT happened to be equipped with just the right instruments to look for a signature of interstellar bacteria.

In February and April 1980, Dayal, collaborating with D.A. Allen, obtained the first spectra of a source known as GC-IRS7 located near the centre of the Galaxy, which showed a broad absorption feature centred at about 3.4 micrometres. When Dayal sent his spectrum to us we found that it agreed in a general way with the spectrum of a bacterium that we had found in the pub-

81

lished literature. But at this time neither the wavelength definition of the astronomical spectrum nor the bacterial spectrum (before Al-Mufti's became available) was good enough to make a compelling case for interstellar bacteria. However, even from these early observations we were able to check that the overwhelming bulk of interstellar dust must have a complex organic composition. Until then the infrared data showing organic polymers in space had related only to localised dust clouds such as Trapezium nebula, which we discussed in Chapter 5. Organic solid particle absorption on a wider scale was a possibility then, but it could not yet be proved.

The observations that were to mark a crucial turning point in this entire story were carried out by Dayal Wickramasinghe and D.A. Allen at the AAT in May 1981. The new observations were of a far superior quality, because a new generation of spectrometers was used. Dayal sent us his raw data by fax to compare with our brand new laboratory spectra which had been obtained just months earlier by Al-Mufti. We were able to overlay the astronomical spectrum over the detailed predictions of the bacterial model of cosmic dust, as can be seen in Fig. 6.2.

We were told by a number of chemists with experience of infrared spectroscopy that a curve like that of Fig. 6.2 can easily be obtained from a mixture of nonbiologically derived organic materials. Since by now we had examined without success literally hundreds of infrared spectra of individual organic compounds, we did not immediately believe this claim. Consequently, we asked the chemists in question to produce an explicit example of such a mixture. But it never was, with the exception perhaps that some expensive laboratory experiments, involving a carefully controlled irradiation of inorganic mixtures, have produced unidentified or poorly specified organic residues that may possess the desired infrared spectral properties to varying degrees.

It cannot be disputed of course that the basic molecular units that cause absorptions such as in Figure 6.2 could be formed by

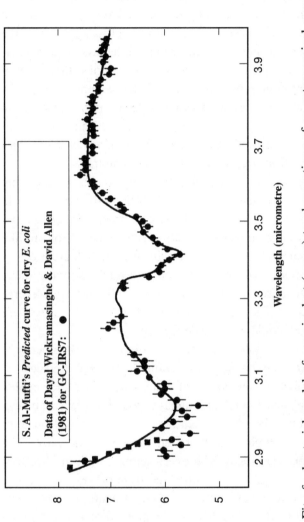

Figure 6.2 Fit of a bacterial model of cosmic dust (curve) to observations of an astronomical source called GC-IRS7 located near the centre of the Galaxy. The points are the very first measurements of this source made in 1981 at the AAT by Dayal Wickramasinghe and David Allen. The impressive fit shown here was argued to prove the predominance of organic particles resembling freeze-dried bacteria in interstellar dust. Later observations of this source and newer data on the absorption properties of dessicated bacteria have not substantially altered this important conclusion.

processes that do not involve biology. The situation is analogous to individual bars of a musical composition that add up to make a sonata. The components of a composition need to be put together in a highly specific way in order to get the final pleasing result. If the functional molecular units in a bacterium can be produced synthetically and mixed together in the correct proportions, the bacterial spectrum can of course be reproduced. What cannot be disputed, however, is that the observations of Figure 6.2 imply that one in three carbon atoms in interstellar space is tied up in the form of such a mixture of functional units that behave in a way that *cannot* be distinguished from freeze-dried bacterial matter.

Furthermore, the observational data in Figure 6.1 implies that this material is distributed in the form of particles that are similar in their sizes and internal structure to hollow freeze-dried bacteria. Are we therefore to assume that we are seeing evidence for life-like particles that must not be connected with life? A hologram of my friend on the street, and not my friend himself? Has reality been transformed into a dream? This seems to me an incredibly perverse position to adopt. The principle of Ockham's razor if it were correctly applied would surely rule this out.

There are major problems in trying to discover what process or processes other than biology are available to lead to such an exceedingly efficient conversion of inorganic matter to complex organics and on such a colossal scale as is found. As far as one can see there are none. Gas phase reactions are woefully inadequate to produce even the complement of organic molecules that exist in the gas phase interstellar clouds. The overwhelming bulk of terrestrial organics that we see around us is derived from biogenic processes. So we are forced to conclude that it is only a matter of philosophical and cultural predeliction that can direct us to seek a solution to this problem from the realm of the impossible, which is what many are still prone to do.

The correspondences seen in Fig. 6.1 and Fig. 6.2 are surely not all that is there. Agreements with the data emerged in whichever direction we cared to look. The spectrum of the Trapezium nebula, which was our starting point of organic dust models, now produced a perfect agreement with a biological model, that combined purely carbonaceous microorganisms with a class of common algae called diatoms that included siliceous biopolymers.

We referred earlier to the story of the interstellar absorption band at the ultraviolet wavelength of 2,175 Angstroms and our incorrect assignment of this feature to graphite. No sooner had we embarked on the organic polymer trail than we began to attribute this absorption to the effect of aromatic carbon-ring structures. Our paper entitled, 'Identification of the λ2,200A interstellar feature' published in *Nature*, November 1977, makes this assignment for the first time in no uncertain terms. From 1980 onwards other infrared observations accumulated that also had a bearing on aromatic molecules in interstellar space. In the mid-1980s, groups of astronomers both in the USA and France independently concluded that certain infrared emission bands occurring widely in the Galaxy and in extragalactic sources are due to clusters of aromatic molecules. The molecules absorb ultraviolet starlight, get heated for very brief intervals of time, and re-emit radiation over certain infrared spectral bands, including one centred at 3.28 micrometres. Needless to say, such molecules are part and parcel of biology, and their occurrence in interstellar space is readily understood as arising from the break-up of bacterial cells. Hoyle and I showed in 1988 (in a paper presented at an European Space Agency conference in Salamanca) that galactic infrared emissions at 3.28 micrometres and other infrared wavelengths, *combined* with extinction at 2,175A can be elegantly explained on the basis of an ensemble of biologically generated organic molecules. Competing nonbiological models and explanations have of course been offered, but all such models are fraught with serious problems, to say the least. No fit as good as the biological model has been presented to date.

A few years ago we unearthed yet another property of biological pigments such as chlorophylls, a property that clearly shows up in the astronomy. Many biological pigments are known to fluoresce, in the fashion of pigments in glow worms. They absorb blue and ultraviolet radiation and fluoresce over a characteristic waveband in the red part of the spectrum. For some years astronomers have been detecting a broad emission features of interstellar dust over the waveband 5,500–8,000 Angstroms. In one instance an entire galaxy M82 is surrounded by dust that is fluorescing in this waveband. It is possible that we are seeing here the effects of flourescence of biological material on a galactic scale.

The facts are unassailable: interstellar dust particles possess spectroscopic and scattering properties over a wide range of wavelengths that make them indistinguishable from bacterial particles. In other words from spectroscopic measurements one cannot say the difference between dust in interstellar space and desiccated bacteria in the laboratory. Moreover, there is a truly vast quantity of such organic dust in the Galaxy, amounting to the involvement of one in every three carbon atoms in the interstellar medium, or a total mass of some 5 million times the mass of the Sun. Much the simplest way to produce such a vast quantity of small organic particles everywhere of the sizes of bacteria is from a bacterial template. It is not from any known form of abiotic process.

Reference

HOYLE, F. & WICKRAMASINGHE, N.C. (2000). *Astronomical Origins of Life: Steps towards panspermia* Kluwer Academic Press.

7. Comets, the Cosmic Dragons

All truths wait in all things,
They neither hasten their own delivery nor resist it . . .
Walt Whitman *(1819–1892)*

A little over 4 billion years ago the Earth was in its final stages of formation. The rocky outer crust, the mantle below and the metallic core were all in place, but the final touches to a magnificent sculpture were still to be finished. At this stage the Earth probably turned on its axis a little faster than once in 24 hours, and the sky was frequently scribed with the majestic trails of comets. Many a comet would have crashed on to the rocky surface of the young planet and in the process brought home a rich bounty of volatile materials such as water and carbon dioxide that went to form the Earth's oceans and atmosphere. Comets cratered and scarred the surface as they struck, but they also eventually transformed a barren lunar-like landscape of the primitive Earth into a planet with water and blue skies, creating a congenial habitat for life.

We mentioned earlier that by what seemed a remarkable act of providence the very oldest evidence of terrestrial life coincides exactly with the moment when cometary collisions had dwindled to a mere trickle. This was a time between 3.83 and 4 billion years ago when the last phase of the accumulation of the planet from cometary material was complete. The only rational way to interpret these facts is on the assumption that life was added to the Earth by comets, and that cometary life simply took root on the planet just as soon as the surface conditions became congenial. It cannot be said too often that there is no evidence whatsoever for any traces of a geological epoch in which a primordial organic soup was supposed to have existed. Nor is there evidence of prebiotic chemical evolution that preceded that appearance of the oldest life on Earth.

On this picture it would be natural to regard comets as the bringers of life to the Earth, and more generally as the carriers and distributors of life and of life's genetic heritage throughout the cosmos. Hoyle and I arrived at this position in 1981 when evidence for bacteria or bacterial type dust in interstellar space seemed to us to have been established beyond reasonable doubt.

Having disposed of the time-hallowed paradigm of the primordial soup, turning to comets for the source of life was a double transgression in the eyes of those who regarded themselves as the custodians of science. A connection between comets and human affairs had been deeply ingrained in many diverse cultures throughout the world and the ancient fear of comets appears to be almost universal. The attempts by Hoyle and myself to link comets to the origins of terrestrial life seemed too much like the revival of an ancient superstition, and so were met with general disapproval.

Ancient ideas on comets fall into two categories. They were either thought of as astronomical objects on a par with the planets, or they were regarded as meteorological phenomena – incandescent vapours that existed only in the atmosphere. In either case the focus was not on the primary phenomenon itself but on the effects that it generated. Aristotle who argued that comets are firy meteors probably came nearest the mark. He further claimed that a cometary apparition ushered in a range of terrestrial effects – stones falling from the sky, storms, droughts, tidal waves and even earthquakes. Once again he seems to have been right. The belief that comets heralded unpleasant effects that affected our lives led them to be considered as omens. Thus Shakespeare wrote:

Comets, the Cosmic Dragons

When beggars die there are no comets seen;
The heavens themselves blaze forth the death of princes.

Undoubtedly these early ideas concerning comets must have been a complex mixture of plain facts and inferences from observing certain coincidences, some of which were perhaps meaningful, others not. Poetic licence may have also played a role, as well as superstition and fear, especially as the nature of the cometary phenomenon remained totally unknown at the time.

The periodicity of cometary appearances were first noted by the English astronomer Edmund Halley (1656–1742). He observed close similarities between the orbital paths of three comets that appeared in 1531, 1607 and 1682 and inferred that it was the same comet making repeated circuits round the sun. By applying Newton's recently propounded theory of gravitation, the French astronomer A.-C. Clairaut (1713–65) was able to predict that the same comet would reappear in 1759, and it surely did. This event of verification provided a convincing vindication of Newton's theory of gravitation. Studies of Newton and Halley left little doubt that cometary bodies are substantially smaller than planets and that they pursued eccentric elliptical orbits around the Sun, this behaviour being a consequence of the inverse square law of gravitation.

Despite all the sophistication of seventeenth-century Newtonian mechanics, and the general emergence of a mechanistic, Cartesian world view, the need for cometary divination was not removed. Newton himself noted that comets move in chaotic elliptic orbits which are randomly oriented with respect to the plane of orbits of the planets. He was therefore acutely aware of the dangers of comets colliding with planets, but he took solace in the belief that Providence or a benevolent God would have arranged matters in such a way that we would be spared from devastating collisions except perhaps on the rarest of occasions. The latter too he thought was important in a positive sense of providing a periodic refuelling of the Earth and planets.

In the late nineteenth and early twentieth centuries studies of comets were confined to amateurs and a small group of committed professional astronomers and physicists. Cometary science was regarded as being largely irrelevant to the understanding of the nature of stars, planets and the largescale structure of the Universe. The dangers to terrestrial life, that were dimly perceived by Newton and expressed in his correspondence, were largely forgotten, or more likely ignored. Comets were, on account of their small masses, considered as inconsequential astronomical objects. Whilst an individual comet measuring some 10 kilometres across might be called small in relation to other astronomical bodies, a 100 billion of them, which still lurk in the outermost reaches of the solar system are not inconsequential and by no means negligible in their effects. Particularly so if comets harbour the stuff of life, as we shall see in this chapter, and if comets colliding with the Earth have destroyed life periodically, as we shall see in the next chapter.

As a comet approaches the Sun in its orbit, solar heat begins to vaporise the surface material. The liberated gases surrounding the nucleus soon begin to glow, producing a fuzzy coma typically extending out to half a million kilometres, a dimension that compares with the size of the Sun. The ultraviolet radiation from the Sun and the outward moving gas from the solar surface then blows the luminous gas and dust in the comet's coma into a tail that can itself extend to lengths of 10 to 100 million kilometres. The tail of a comet often splits into two – one wispy thin tail made of gas, and a broader, gracefully curving fan-shaped tail made of dust, the dust tail often fanning out over an angle of 10–20 degrees across the sky. The result, particularly for a large comet, is one of the most spectacular sights visible in the night sky. Recent comets that have graced our skies in this way include Comet Halley, Comet Hyakutake and Comet Hale-Bopp.

When we concluded that interstellar dust carried the stuff of life, the connection with comets seemed more or less well defined. In the 1970s comets were thought of as uninteresting lumps of dirty ice – a model that was invented by the American astronomer Fred Whipple, and one that had come to be regarded as gospel truth in the scientific community for over two decades. All that was soon to change with the arrival of new observational evidence, but the final abandonment of the Whipple dirty-ice cometary paradigm was very much harder than one might have imagined.

Whipple's dirty ice conglomerate or dirty iceberg model came into vogue after the discovery, in cometary comae and tails, of a number of emission bands corresponding to many atomic species and simple molecules such as water (H_2O), carbon dioxide (CO_2), carbon monoxide (CO), hydrogen H, sulphur S, C_2, C_3, cyanogen (CN) methylidyne (CH) the hydroxyl radical (OH) and the amino group (NH_2). It seemed reasonable then to suppose that a mixture of water, ammonia and methane served as a source of the parent molecules. Cometary tails had been shown to have the spectral characteristics of sunlight scattered from small particles about a micron in size. These solid particles were thought to be mixed in with the more volatile material that makes up the bulk of the nucleus of the comet. As the volatile gaseous material is expelled from the nucleus, it carries the dust component along with it. In the 1970s the dust in the tails of comets was thought to be made up of mineral silicates, similar to a component of household dust. A crude representation of the Whipple dirty iceberg comet was a 10 kilometre-wide chunk of frozen water, methane and ammonia with an admixture of silicate dust and other molecules sitting lightly within cavities of an icy matrix.

As we progressed through the early 1970s, radio astronomers were discovering molecules of formaldehyde (H_2CO), methyl cyanide (CH_3CN) and hydrogen cyanide (HCN) in the gases emanating from comets. The new data did not fit neatly into the reigning cometary paradigm, and it certainly made one think. At this time Kopal introduced me to his fellow Czech compatriot V. Vanysek, Professor of Astronomy at Charles University in Prague. Vanysek had been studying the spectra of comets for some time, and had

91

himself begun to feel uncomfortable about the Whipple dirty iceberg model. Together with Vanysek I proposed a radically new model for dust in comets. We suggested that the bulk of the cometary nucleus was made of refractory organic material, with organic polymers breaking loose at the surface and further fragmenting into smaller molecular units as the comet came close enough to the sun. The infrared spectrum of Comet Kohoutek obtained in 1973 seemed to us to provide compelling evidence for this model.

There was a lapse of several years before Hoyle and I began to contemplate much deeper isues that related to comets. What if the comet dust was not just composed of nonbiological organic polymers, but intact bacteria and their degradation products? What if there existed a connection between interstellar dust, which we had argued was biogenically derived, and the dust seen in the magnificent tails of comets? This would instantly elevate the status of comets from the insignificant objects that they had been thought to be to the most interesting and important objects in the entire Universe.

We saw in Chapter 2 that the formation of the comets in the early solar system involved the inclusion of interstellar dust grains from molecular clouds such as are seen in the Orion nebula, and there is ample evidence nowadays to support this contention. If the original comet reservoir, the Oort cloud of comets, had incorporated even as little as a trillionth of interstellar dust grains that were in the form of viable bacteria – bacteria that survived a 100 million year sojourn in deep space, and this is surely an exceedingly modest requirement – there would be hundreds of viable bacterial cells per comet at the outset.

Each one of the 100 billion comets in the Oort Cloud would then be assumed to start life with hundreds of viable bacterial cells in a freeze-dried dormant state. Trapped as they would be within a lump of frozen ice, this legacy of cosmic life would come to nothing but for the next important step that must inevitably occur. The falling together of cometary bolides and dust to become a full-blown comet would have involved inclusions of molecular fragments called radicals that can re-combine to release chemical energy, and also radioactive elements formed in a nearby supernova that could release energy in the processes of radioactive decay. Both radio-active heat sources and chemical energy released slowly would thus have led to a phase of internal melting.

Fledgling comets could thus have harboured a modest sprinkling of viable microbes in warm watery pools that developed at their centres. We could think of each comet as a gigantic culture medium containing, to start with, a few viable cells in the midst of water, organic nutrients, and all the necessary inorganic salts needed for rapid microbial growth. The volume of this culture medium would be vastly bigger than anything that could be imagined in a bio-chemist's laboratory, bigger than the Empire State Building, bigger than the largest orbiting space station that is planned for the foreseeable future. For as long as an interior region of a comet remains melted, these biologically favourable conditions within an individual comet seem inescapable.

At the time we were discussing these ideas in the early 1980s Max K. Wallis, whose interest was in interplanetary and cometary plasmas (ionised gas in the presence of a magnetic field), had just turned his attention to the exploration of Halley's comet that was being planned by several international space agencies. Max Wallis was a Co-investigator on ESA's Giotto mission to Comet Halley from 1980 to 1986, and during this period emerged as one of the leading cometary theorists in Europe. Hoyle and I had the benefit of many conversations with Wallis at the time when our heretical ideas of cometary life were taking shape. In 1980 Wallis turned his attention to the internal structure of comets. Using mathematical models of heat transfer he confirmed, by means of detailed calcula-tions, that the interior of a large enough comet heated by radio-active heat sources could remain liquid for millions of years beneath an insulating frozen layer of ice. This work was published in *Nature* (Volume 284, 1980, page 43), and it should be mentioned in passing that the journal editors had excised a reference to the prior work in this area by Hoyle and myself, showing once again that the main science journals and scientific establishments regarded any reference to panspermia as a heinous crime.

The length of time for which the interior liquid state is maintained within a comet depends on the radioactive heat sources that are considered, Uranium, Thorium and ^{26}Al being possibilities, and also the overall radius of the comet. An outer shell of thickness of about a kilometre in a 10 kilometre radius comet could provide enough insulation to preserve a warm watery interior for a long enough time for a replicating bacterium to be able to swamp and colonise the bulk of a comet's volume. If one imagines a single viable anaerobic bacterium placed within the liquid interior of a comet, it would divide to yield two offspring in two to three hours. These then divide sequentially every two hours, two becoming four, four becoming eight, eight becoming 16 and so on. With 40 doublings and with continued access to nutrient, the culture expands to the size of a marble in four days. With 80 doublings the culture would grow to the size of a village pond in eight days, and with only 120 doublings an entire cometary core would be converted to bacterial matter in a mere 12 days. The precise timescales in this example are underestimates of course because we assumed instant access to nutrients and energy at all times. A more realistic timescale of cometary transformation into biomaterial may well be in the order of hundreds of years, but this is well under the timescales that are surely available for the preservation of liquid interiors in comets. And so it follows that the 100 billion comets in the early history of the solar system would have provided the most powerful setting for amplifying cosmic bacteria – for regenerating and perpetuating the legacy of cosmic life. With each one of 100 billion sun-like stars in our Galaxy providing similar conditions, the scheme for amplifying and maintaining cosmic biology on a cosmic scale is impressively powerful.

Well, how one might ask, could this seemingly crazy idea be tested? The first opportunity arose in the run-up to the 1986 return of Halley's Comet, a matter to which we already briefly referred. It would seem entirely appropriate that an important paradigm shift relating to the composition of comets came to be associated with this truly historic comet. This was the comet that Edmund Halley had studied in 1531, 1607 and 1682 and it was these studies that led to the realisation that comets pursued elliptical orbits around the Sun. Halley's Comet has an average orbital period of about 76 years, and its last perihelion passage or closest approach to the Sun took place in February 1986. Needless to say this event was unique in the scientific history of recent times.

Ahead of the European spacecraft Giotto reaching Comet Halley, Hoyle and I published a paper entitled 'Some Predictions of Comet Halley' early in 1986. Here we made predictions of what the cameras aboard Giotto would see at closest approach to the nucleus of the comet. We argued that the surface would look dark, roughened and generally coal-like, the result of the prolonged weathering of biological polymers. That is precisely what was found, a surface that was described as 'black as coal'. The architects of the Giotto project paid a heavy price for their blind faith in a wrong theory. They expected to find a surface as bright as a snowfield and accordingly trained their cameras to focus on the points of greatest intensity. The dark surface of the comet therefore caused a problem in the pictures that were initially beamed, leading to the disappointment of many millions who watched the Giotto-Halley encounter on their TV sets. Later processing of the images revealed a giant peanut-shaped image of the nucleus of Halley's comet measuring some $16 \times 8 \times 7$ kilometres – the first comet nucleus that had ever been directly seen at close quarters.

Although the millions of dollars spent on Giotto did eventually lead to important results, the first strong support for the cometary life hypothesis followed from observations made by Dayal Wickramasinghe and David Allen at the Anglo Australian

Telescope on 31 March 1986. When we received a fax of their new data as it came out of the telescope, we were ecstatic. It was yet another triumph for our life from space theories. The infrared spectrum of dust from Comet Halley matched precisely the laboratory spectrum of heated bacterial grains. Comet Halley was expelling bacterial type dust at the rate of a million or more tonnes a day. And this is what Comet Halley kept on doing for almost as long as it remained within observational range. An independent analysis of dust impacting on mass spectrometers aboard the spacecraft Giotto also led to a complex organic composition, a composition that was fully consistent with the biological hypothesis. Broadly similar conclusions have been shown to be valid for other comets as well, in particular Comet Hyakutake and Comet Hale-Bopp. Thus, one could conclude that cometary particles just like the interstellar dust are indeed indistinguishable from bacteria. The simplest conclusion would be that they are indeed bacteria, particules derived from the amplification of cosmic biology within comets. There is an old saying that if something looks like a duck and quacks like a duck, then it had better be a duck. Cometary dust now had all the signs that made them look like bacteria and 'quack' like bacteria, so commonsense dictates that they most probably *are* bacteria. Of course the obstinate critics of panspermia, just like the supporters of the old Ptolemaic world view would invent new epicycles, new explanations of how life-like organics could arise from processes unconnected with life. But their days are surely numbered and all the facts seem to be moving inexorably in the direction of cometary panspermia – the dispersal of life by comets.

Reference

HOYLE, F. & WICKRAMASINGHE, N.C. (1981). *Living Comets*, University College Cardiff Press.

8. Dragons Fall from the Skies

Like a blazing comet, seems destined either to inspire
with fresh life and vigour, or to scorch up and destroy
the shrinking inhabitants of the Earth . . .
Thomas Robert Malthus *(1766–1834)*

Comets were feared without exception by most ancient cultures throughout the world. They were depicted in art, sculpture and literature as evil omens, foreboding pestilence, disaster and even death. To us living at the dawn of the twenty-first century these ancient beliefs seem irrational. We put them down to ignorance, even stupidity of ancient Man. We shall show in this chapter that such an outright condemnation of our ancestors cannot by any means be justified.

The meteors one sees streaking across the night sky on a moonless night are minute pieces from a comet. Each meteor starts its life as a tiny object between a pinhead and pebble in size, and the glowing streak of light is none other than the incandescent trail of hot gas that it leaves behind as it burns. Occasionally an incoming meteoritic missile will be so big that not all of it burns up and a piece may even land intact. Very rarely a large meteorite has been recorded to crash through a roof, but such events are so infrequent that we can almost ignore such hazards in our everyday lives. But whether this comforting situation was the same as it is today in past times, remains to be seen.

Comets from which meteors and meteorites are mostly derived are by far the most spectacular objects in the night sky. We have already noticed that comets pursue elliptical orbits around the Sun, and that a 100 billion or so cometary bodies – objects left over from the formation of Uranus and Neptune – still surround the planetary system in the form of a gigantic shell of comets. The effect of passing stars on this shell is to deflect individual comets and make them plunge in highly elongated elliptical orbits into the inner regions of the solar system. When such a deflected comet reaches a point about three times the radius of the Earth's orbit from the Sun, evaporating gas and dust form a diffuse glowing nebulous coma from which cometary tails stretching over millions of kilometres are derived. A comet's first incursion into the inner regions of the solar system takes place along an orbit that has a period in the range of hundreds of thousands of years or even longer. After a number of transits through the inner regions of the solar system a long-period comet becomes eventually transformed into a comet with a much shorter period due to the build-up of gravitational interactions with the giant planets Jupiter and Saturn. The shortest-period comets have periods of three to four years. Such comets go out nearly as far as Jupiter in their elliptical orbits.

The source of the magnificent splendour of a comet is its nucleus, which we saw, in the last chapter, is a chunk of frozen organics and water ice measuring, on average, some 10 kilometres across. Comet Halley has a dimension roughly about this size. A class of comet that has recently come to the fore is that of the 'giant comet' with a typical radius in the range 30–200 kilometres. The existence of giant comets was predicted by British astronomers Victor Clube and Bill Napier but remained only as a conjecture until the discovery of the object known as Chiron in 1977. At first it appeared that Chiron was just another asteroid moving in a somewhat peculiar orbit that lay between the orbits of Saturn and

Uranus. In 1989, however, Chiron's cometary character was revealed when it was found to brighten enormously and develop an extended cometary coma. Chiron was the first giant comet to be discovered. Subsequently more than two dozen similar objects have been found, mostly present in the outer regions of the solar system near the orbits of Neptune and Pluto.

A member of the class of giant comets to grace our skies a few years ago was Comet Hale-Bopp. This comet had an estimated diameter of 40 kilometres (25 miles) and came towards the lower end of the size range expected for giant comets. Hale-Bopp made its closest approach to the Sun on 1 April 1997, and it is due to return again to the inner solar system in the year AD4377. The effect of Jupiter on comets is markedly evident in the case of Comet Hale-Bopp. The orbital period of revolution, as it came in from the depths of the solar system, was 4,200 years, but its passage past Jupiter at the modest distance of 65 million miles reduced this period to 2,380 years.

The first half billion years of the Earth's history was riddled with comet impacts. The collisions at this epoch represented the final stages of the formation of the Earth as we presently recognise it. Although this initial impact epoch, known as the Hadean epoch, came to an end about 4 billion years ago, collisions with comets, asteroids and fragments of comets did not come to a complete halt.

So what were the direct physical effects that resulted from the collision of comets and cometary fragments with our planet? Many of the planets and satellites of our solar system that have been studied and photographed at close quarters have revealed heavily cratered and pockmarked surfaces. In general, planets with atmospheres show less in the way of surviving scars, for the simple reason that impact scars are relatively quickly smoothed over by surface weathering. But a close inspection of the Earth shows abundant evidence of impacts. About 150 impact scars have been identified to date, including giant craters buried under sediment in

Iowa, and a crater that lies beneath most of Chesapeake Bay in the United States.

There is now little doubt that the extinction of the dinosaurs who ruled the Earth for millions of years, and of over 75 per cent of all genera of plant, animal and microbial life occurred 65 million years ago as a result of a comet impact. A process of this kind was first suggested by Hoyle and myself as early as 1978. In the winter of 1977 we were already discussing the role of hollow organic dust with biological connotations as models for the dust particles that pervade the recesses between stars. We now started to argue that similar particles will be incorporated in comets and moreover that the Earth will encounter such comets on the average of once in about 100 million years. This would be the mean interval between two direct strikes of a full-blown comet nucleus. In our paper entitled 'Comets, Ice Ages and Ecological Catastrophes' in *Astrophysics and Space Science* (vol. 53, page 523–526, 1978) we wrote that 'Although more dramatic and seemingly more devastating, a direct hit would not necessarily be as far-reaching in its effects as the addition of 10^{14} grammes of small grains . . . would be.' We still maintain that a protracted episode of atmospheric dusting by comets, which would also occur once in about 100 million years, would be by far the most important process in precipitating a major ecological catastrophe such as occurred 65 million years ago. There is some evidence to suggest that the extinction of species at this time culminating in a single major event occurred over a protracted period of some 100,000 years. A veil of dust in the stratosphere would dim sunlight to a level that photosynthesis by microscopic plankton in the ocean at the bottom of the food chain of all animals would fall to a very low rate. Leaves would wither from trees, leading eventually to extinctions of large browsing animals, as happened in the case of the dinosaurs 65 million years ago.

A very similar process connecting the impacts of comets with the extinctions of species was being discussed in parallel by Bill Napier and Victor Clube, and published in a paper that appeared in *Nature* (vol. 282, 29 Nobember 1979).

Two years after our comet theory was published L.W. Alvarez, W. Alvarez, F. Asaro and H.V. Michel discovered an enhancement of the chemical element called iridium in the Earth's sedimentary deposits of 65 million years ago, pointing clearly to the involvement of comets. Their work was duly published in the American journal *Science* (vol. 208, 6 June 1980), but with no reference whatsoever to the earlier papers by Hoyle and myself, or by Napier and Clube. This type of cavalier rejection of priorities in science has become commonplace in modern science, fanned mostly by the desire to secure the maximum amount of public funding for scientific research projects. Normal standards of morality and decent conduct would seem to have gone by the board in this process.

The element iridium is exceedingly rare on the Earth but is much more abundant in comets and meteorites. Thus, whenever we see a layer of terrestrial sediment that is rich in iridium, we can be sure that a comet or comets collided with our planet. Later studies showed that certain amino acids normally found in meteorites, but not so commonly in life, are also distributed in the same sediments, again pointing to the addition of cometary or meteortic material. Much later it was found that the comet collision that coincided with the extinction of the dinosaurs has also left its mark in the form of a crater buried in the Earth's crust near the village of Chicxulub in northern Yucatan.

Other large-scale extinctions of species at earlier epochs have also been found to be associated with enhancements of the same element iridium in geological sediments, so a cometary connection is again believed to follow. Significant peaks in the extinctions of species have been discovered in the geological record at approximately 1.6, 11, 37, 66, 91, 113, 144, 176, 193, 216, 245 and 367 million years ago. Indeed, the entire pattern of mass extinctions of species over the past 400 million years is strongly suggestive of recurrent catastrophic events of external origin with a significant periodocity showing up at about 26 million years. The evidence accords well with ideas discussed by Napier and Clube, suggesting that up and down oscillations of the solar system about the central plane of the Galaxy produces periodic 'galactic tides' that dislodge comets from their normal orbits in the outskirts of the planetary system.

We should not be content to think that dramatic effects of comet collisions are only confined to distant geological epochs. More recent episodes of comet and asteroid impact have surely left their mark, not only as scars on the planet to which we have already referred, but upon our history as we shall see.

To understand what happened it is important to remind ourselves about the nature of comets. We saw earlier that a comet is not a dirty snowball as it was once thought to be. Recent studies have shown that comets must possess exceedingly fragile internal structures. This is because they started their lives with radioactively heated liquid inner cores that subsequently refroze. Re-freezing from outside just like freezing-up of pipes in the winter leads to a cracked and broken-up structure that is very much weaker than a solid chunk of ice. Such a comet is easily split as indeed has been witnessed in many comets in recent times.

On 25 July 2000 a comet called Comet Linear was heading towards its perihelion or closest approach to the Sun. It was a relatively unimpressive object, visible only with a pair of binoculars, until it suddenly flared up and brightened, producing a huge coma of gas, like a cork popping out of a champagne bottle. Then it mysteriously disappeared for several days, but only to reappear again as a much fainter comet. Long exposure photographs revealed now that the comet was split into over 20 separate small pieces with each piece contributing to a much weakened cometary coma. The weakly welded multicracked structure of the comet was obviously torn apart by the force of the evaporated gas that built up huge pressures inside, prior to popping the champagne cork on 26 July.

Comets for the most part stay very far from the range of the inner planets occupying orbits that envelop the entire planetary system at a great distance. Occasionally a comet swings past the giant planet Jupiter. When this happens it is not only deflected in its orbit, so that it no longer returns to the most distant regions from which it came, it can also be shredded into pieces. This happened in the summer of 1992 for the now famous Comet Shoemaker-Levy 9. The comet broke up into a string of 21 pieces due to the tidal forces of Jupiter. In this case the pieces were pulled in to become satellites of Jupiter moving in orbits that slowly spiralled in towards the planet. Over the course of a single week in July 1994 all 21 pieces of Comet Shoemaker-Levy 9 crashed on to Jupiter, one every eight hours or so. The scars left on the face of Jupiter imply an explosive power measured in millions of equivalent Hiroshima-type bombs.

One of the most impressive surviving scars on the Earth is the Barringer Meteorite Crater near Flagstaff, Arizona. This was caused not by a comet but by an iron meteorite some hundred metres across that struck about 20,000–30,000 years ago. The estimated energy of this meteorite impact is close to a thousand times the energy of the atom bomb that hit Hiroshima in 1945. In 1947 the Earth was cratered on a smaller scale by the Sikhote-Aline meteorite, weighing an estimated 70 tonnes. This object flashed in the skies above Russia as a brilliant fireball in February 1947 and broke up at a height of some six kilometres. As the pieces fell, hundreds of craters were formed, with diameters ranging from half a meter to 27 metres. The estimated energy of the impacting object in this case was a little less than one tenth of the energy of the Hiroshima bomb.

The history of civilisation, if it is correctly read, bears witness to the most recent chapter of collisions by comets and fragments of comets, collisions that in effect controlled the fate and progress of mankind. The ideas that I shall describe follow generally the theories advanced by Napier and Clube, but with modifications that were introduced by Hoyle and myself concerning some matters of

103

detail. The precise identity of the giant comet and its original path are left unspecified. The original giant comet itself would have long since ceased to exist, because it fragmented in the manner of Comet Linear or Comet Shoemaker-Levy, forming a meteor stream comprised of billions of individual pieces.

The fragmenting giant comet, call it Comet X, must have weighed of the order of 10,000 million million tonnes, 1,000–10,000 times the mass of Halley's Comet, and measured some 100 kilometres across. It would have become perturbed by Jupiter, say 16,000 years ago, into a periodic orbit that crossed the orbit of the Earth. Within a few cometary revolutions around the Sun the fragmentation would have proceeded sequentially in the style of Comet Linear to produce sequential disintegration into fragments of decreasing sizes, ranging from say 10 kilometres to 100 metres and less. The kilometre-sized pieces could number millions and the 100 metre-sized pieces could number billions. All the pieces would of course remain bunched together within a meteor stream and the evolution of their orbits will be such that the Earth encounters the débris with some well defined periodicity. This average time interval between episodes of bunched collisions is difficult to pin down with any degree of precision. Encounters could start with the initial period of Comet X, but as the meteor stream disperses and evolves the situation becomes more chaotic.

We shall now proceed to chart the progress of Comet X and its progeny through time by attempting to identify crucial events in the history of our civilisation that appear to be connected with possible cometary incursions. In this way we avoid an awkward dynamical problem and work backwards from historical facts to determine the time between successive encounters with the meteor stream of Comet X. Each encounter will be characterised by an epoch of bunched collisions with cometary missiles. During a bunched collision epoch, impacts of objects a few hundred metres across would pose a far greater threat to life and limb than impacts of much larger pieces that would occur far less frequently. For instance, a kilometre-sized object may have collided perhaps only a few times during the 16,000-year timespan which is under discussion. Collisions with smaller fragments, on the other hand, would have occurred with menacing frequency during the bunched collision epochs, each of which may have lasted several decades.

During the past 16,000 years perhaps the most important geological event was the emergence of the Earth from the last ice age. According to our point of view (developed in more detail in Chapter 11) this event would have been triggered by the impacts of cometary fragments. The process of unlocking the planet from its glacial state took place in several stages, as is indicated by the record of temperature and precipitation rate for instance. The first stage may have been caused by a collision with a large cometary fragment. The Earth warms up briefly but then cools again to remain intermittently glaciated for the next 1,500 years, during which minor temperature oscillations may have been caused by a succession of much smaller impacts. The first significant upward jump of temperature occurred at about 9,500BC, due perhaps to the collision of a slightly larger piece of the comet. But even this one collision did not suffice to lift the Earth into a relatively stable interglacial. There was then a sharp plunge back to ice age conditions and several sharp oscillations of temperature followed thereafter, probably dicatated by impacts. It is during this period that many species of large mammals suddenly became extinct in the northern hemisphere, including sabre-toothed cats, mammoths and mastodons. Paleontologists still argue about the cause of this extinction episode, but an impact-mediated climatic change seems the most promising hypothesis at the present time. A decisive emergence of the Earth from the last ice age had to await a major collision event 10,000 years ago. Water that was released due to evaporation from the oceans by the heat of impact was sufficient to restore the greenhouse effect on a very short timescale, thus causing the Earth to pass into a warmer phase.

From this moment in time, at about 8000BC, the history of human civilisation can be said to have begun. At later times there is evidence of fluctuations in the Earth's average surface temperature over timescales of centuries to millennia. Fluctuations above and below the present-day value occur in the general range of three to six degrees F. It is hard to find a purely Earth-based mechanism that could explain this pattern, but again impacts of cometary fragments provide a possible answer. Comets, frequently breaking up in the high atmosphere or in the near-Earth environment, would inevitably produce an increase in the dust loading of the stratosphere, thus enhancing the reflective power of the atmosphere to solar radiation. The result is that less sunlight gets through the atmosphere to heat the surface, more is reflected back into space, so cooling takes place. Thus, it is entirely natural that dips in the average temperature profile or the Earth's surface would be associated with episodes of bunched comet impacts.

On this picture it is clear that several episodes of bunched collisions with pieces of our Comet X must inevitably have occurred over the past 16,000 years. A provisional list of major events possibly connected with such impacts is laid out in Table 2. The far left column gives a notational time separation between the listed events. A mean separation of 1,800 years can be calculated, although this is not particularly meaningful, because some of the events in this calendar near the top may not have been fully charted.

For a typical cometary bolide of radius 40 metres hitting the Earth head-on at a speed of 14 km/s the kinetic energy of impact is equivalent to about 2 MT (megatonne) of the explosive TNT, that is, the equivalent of about 100 bombs of the type that destroyed Hiroshima in 1945. Such an object, if made of ice, would explode at about 30 kilometres above the surface of the Earth without any notable effect on human life. An object of only three times this size, a quarter of a kilometre in diameter, would strike much lower on

Table 8.1 Calendar of salient events probably linked to impacts

Interval	Date	Event
	14,000BC	Giant comet deflected by Jupiter into Earth-crossing orbit Fragments, hierarchically into progressively smaller pieces
1,500 yr		
	12,500–	Minor impact event; sudden melting of glaciers
1,500 yr	11,000BC	Sequence of minor impacts till 11,000BC
1,500 yr	9,500BC	Major impact event: temperature rise
3,700 yr	8,000BC	Second major event: decisive end of ice age
1,500 yr	4,300BC	Copper smelting discovered
	2,800BC	Step pyramid construction in Egypt Collapse of 1st Dynasty in Egypt Flood mythology gains ground
300 yr		
	2,500–	Collapse of Indus Valley civilisation
	2,300BC	End of Old Egyptian Kingdom Pyramid age; tree ring event
~1,200 yr		
	1,350–	Destruction of Jericho;
~1,800 yr	1,100BC	tree ring event
	AD 540	Collapse of Roman Empire Hun Revolution in India Upheaval in Chinese society; tree ring event

the Earth and cause localised destruction on a significant scale, perhaps on the order of a city. A larger cometary bolide with a diameter of a kilometre would have an energy of 100,000 Hiroshima bombs and would cause widespread damage, perhaps on the scale of an entire country. So the size of the incoming missile is the important fact that determines what the eventual outcome of a collision would be. The bigger the object the greater the damage.

An object of about 100 metres across entered the upper atmosphere of the Earth over Tunguska in Siberia in the early hours of 30 June 1908. A great fireball was seen to pass low over the town of Kirensk and came down over a remote part of the Siberian tiaga. The object did not reach the ground but exploded in the atmosphere at a height of about eight kilometres. The brilliant fireball, said to have outshone the Sun, was seen as far as 1,000 kilometres away from its point of descent; and the sound of the explosion was heard at even greater distances. Well documented accounts of what happened on this fateful day are scant. The nearest to an eye-witness report is to be found in the Russian Newspaper *Sibir* in Irkutsk, Siberia of 2 July 1908.

Early in the ninth hour of the morning of June 30, a very unusual phenomenon was observed here. In the village of Nizhne-Karelinsk (200 versts north of Kirensk) in the north-west sky very high above the horizon, the peasants saw a body shining very brightly, indeed too bright for the naked eye, with a blue-white light. It moved vertically downward for about ten minutes. The body was in the shape of a cylindrical pipe. The sky was cloudless, except that low on the horizon, in the same direction as the luminous body, a small black cloud was seen. The weather was hot and dry, and, as the luminous body approached the ground (which was covered by forest), it seemed to be crushed to dust, and in its place a vast cloud of black smoke formed, and a loud explosion, not like thunder, but as if from the avalance of large stones or from heavy gunfare was heard. All the buildings shook, and at the same time a forked tongue of flame burst upward through the cloud. All the inhabitants of the village ran into the street in terror. Old women wept, everyone throught that the end of the world was upon them. . . .

The immense blast wave felled trees over a distance of some 40 or 50 kilometres, and the heat from the fireball charred tree trunks for distances of up to 15 kilometres from the centre of impact. This scene of devastation, first photographed in 1927, is shown in Plate 8.1. This was real. Estimates of the total energy of the impacting object range from 13–30 megatonnes of TNT, equivalent to the explosive power 650–1,500 Hiroshima bombs.

Collisions of the Tunguska type, and others on a much grander scale, must surely have occurred sporadically as well as recurrently throughout our history and prehistory for the past 16,000 years. With bunched collision episodes separated by intervals of 300–3,700 years as seen in Table 8.2, long periods of relative quiet can be seen to be interspersed by tens or hundreds of years of fierce bombardment.

The odds of a person being killed in a Tunguska-type event is calculated by dividing the lethal area around a point of impact, say

Plate 8.1   Photograph of Tunguska 1908 site taken in 1927 showing scene of devastation following comet explosion.

an area of 5,000 square kilometres, by the total area of the Earth's surface, about 100 million square kilometres, giving an odds of one in 20,000. If such impacts took place at a rate of one per year, the odds of perishing from a Tunguska-type impact would be roughly the same as it is today for death from an accident on our motorways, that is chance of about 10,000:1. But during an episode of bunched impacts with 100 impacts per year the chance over a lifetime of 30 years, say, would be about 30 per cent. By far the most important consequence of such events is that over a bunched collision epoch, occupying say a century, during a bad spell, so to speak, more than one population centre in three would be totally destroyed. Every survivor would have observed what appeared like fire pouring down from the sky from a neighbouring conurbation, perhaps tens of kilometres away. Medieval depictions of scenes, such as reproduced in Plate 8.2 of which there are many, are clearly understood on this basis. It would be hard to imagine that experiences such as these would not have made an indelible impression on the minds of our ancestors. Their effect on the evolution of belief systems would have been profound.

At the dawn of civilisation some 10,000 years ago, cometary activity, including dust production, following the break-up of Comet X, would have been conspicuously intense. For much of the year the entire zodiac, the band across the sky where the planets are seen, may have glowed from the sunlight that was diffusely scattered by cometary dust particles. Sightings of comets ferociously breaking up, rising and setting, displaying their brilliant dust tails would have been exceedingly common in the ancient skies. There can be little doubt that myths and legends would have evolved in response to such experiences, experiences that must surely have been shared by many nomadic tribes scattered widely across our planet. With the power of the sky so manifestly clear it is easy to understand why the gods of most early societies were placed in the skies. One could also understand why such sky-centred religious systems often depicted celestial combat myths in

110

Plate 8.2 Medieval mural showing effects of cosmic missile attack. (Courtesy of W.M. Napier)

which winged serpents or dragons battle for supremacy of the sky, until the vanquished eventually crash down on to the Earth. Ancient dragon myths were surely born in this way.

Scarcely a millennium after the end of the last ice age our ancestors had discovered agriculture and farming, and thereafter began to relinquish their customary lifestyle as hunters and food-gatherers. As the first human settlements – villages and towns – began to appear on the surface of our planet, assaults by cosmic missiles of the Tunguska-type and super-Tunguska-type would have occurred with unrelenting ferocity.

The rises and falls of civilisations, the ascendancy and decline of empires, that punctuate human history over the last 10,000 years, can be explained elegantly on the basis of periodic assaults from the skies. The falls of civilisations occur dramatically during the shorter bad periods and the ascents to power and glory would be sustained over the longer periods of calm.

An important connection between civilisation and cometary encounters was pointed out by Hoyle. As far as human progress goes, perhaps the first crucially important development that may be related to comet impacts is connected with the discovery of metal smelting. This was indeed a most remarkable discovery that led eventually to the use of metals for weapons, tools and equipment. It marked a turning point in the fortunes of man. The idea of converting a piece of stone into malleable metal could surely not have occurred to anyone as an abstract concept. The discovery of metal smelting must therefore have been an unplanned accident, but then the problem is to understand how the same accident happened at the same time in widely separated locations on the Earth. Archaeologists tell us that copper was being used for making tools and utensils at a date close to 4300BC. The first recorded use of pure copper is to be found in Eastern Anatolia, but very quickly it spreads across the world. The indication is that the relevant natural accident that produced copper smelting had to be repeated in several locations almost simultaneously. This was almost certainly the multiple impacts of cometary bolides. Events of the Tunguska-type could have started forest fires on a huge scale. Beneath the intense heat of glowing charcoal, rocks containing appropriate metallic ores would become naturally smelted. Our nomadic ancestors did not need any exceptional powers of observation to come across sites of smouldering forest fires similar to those that raged in Tunguska in 1908. They would only have had to pick up pieces of the smelted copper to discover that they could be beaten and flattened to yield artifacts that served their needs. This marked the beginning of the Copper Stone Age as it is sometimes called, which led to the Bronze Age 1,000 years later. This was an important milestone in human progress.

Flood mythology epitomised in the story of Noah's Ark pervades many ancient cultures. Although Archbishop Ussher's chronology dates this event at 2349BC, more reliable earlier dates for a Deluge at around 2800BC are strongly suggested according to the Chinese annals. W. Bruce Basse has put together evidence from a variety of sources suggesting a major global cataclysm at around this time. There is evidence pointing to huge movements of population and changes in settlement systems the world over. Basse points to intriguing evidence of the rapid dispersal of four large language groups, including the spread of Indo-European languages throughout Europe at about this time. Archaeological evidence for a 2800BC event, though it exists, appears to be scant, and it is interesting to note that this time also coincides with the end of the First Dynasty in Egypt. It is also the time when pyramid building commenced in a modest way with the construction of 'step pyramids' that eventually evolved into the more impressive smooth-faced structures in later years.

A better documented episode or episodes of bunched collisions span the period 2500–2300BC. Just before this time several great civilisations and long-lived dynasties are known to have flourished, both in ancient Egypt and in the Indus Valley of North India. The ruined city of Mohenjo-daro in northern Pakistan appears to have been the site of a pre-Aryan civilisation that was perhaps more advanced than that of Egypt. It had flourished for over a millennium, but suddenly and dramatically collapsed. Aryan invastions from the West could have produced a slow erosion of the old Empire, but not a seemingly cataclysmic fall. Likewise, seasonal flooding of the Indus valley could have had a slow cumulative effect over many centuries though not a sudden one. A far more dramatic catastrophe could have arisen from tidal waves and tsunamis that follow quite naturally when cometary fragments crash into the sea. There has been recent evidence to suggest that a dust layer and a burnt surface horizon apparently caused by an air blast exists in archaeological sites in Northern Syria, sites dated at about 2350BC. This also accords fully with the present point of view.

Perhaps the strongest evidence of an episode of cometary impacts is to be found in the deserts of Egypt. Following the unification of upper and lower Egypt by King Menes around 3100BC, a desert empire flourished, reaching heights of glory, through a succession of dynasties of the so-called Old Kingdom of Egypt, until its eventual collapse at about 2160BC. Pyramid texts describe a prolonged period of turmoil preceeding collapse.

The construction of the three most famous Giza pyramids began with Snefru's son Khufu who built the Great Pyramid around 2500BC. This stupendous structure that never fails to impress the tourist has a base that covers some 13 acres and a height of over 450 feet. The geometrical precision of the construction as well as the exact alignments of its faces to cardinal points (NSEW) is remarkable to say the least. Two other major pyramids were built at Giza over the next two centuries by Khufu's son Khafre and his successor Menkaure.

The three pyramids of Giza taken together can be seen to possess an overall plan that is connected in some way with the stars. Robert Bauval first pointed out that aerial photographs of the three Giza pyramids showed an astounding similarity to the disposition of the three brightest stars in Orion's belt. The agreement, although not exact, includes the relative distances apart of the three pyramids, their sizes in relation to the stellar brightness and a kink in the lines joining the three pyramids that matches a similar kink in the lines joining the stars on the sky. The preoccupation of the pyramid architects with Orion belt stars shows up even more strikingly in the alignment of a channel through the body of the Great Pyramid that connects the king's chamber to the outside world. In 1964 astronomers Virginia Trimble and A. Badway calculated that, at about the time the pyramid was built, this channel would have pointed precisely in the direction of the brightest star in Orion's belt when the star rose to its highest point in the sky. Yet another channel in the pyramid was shown to point to the star Thuban in the constellation of Draconis. This star in the Pyramid Age would have been the pole star around which all other stars would have appeared to revolve.

For any particular meteor stream that the Earth crosses, meteors would always appear to radiate from a fixed point amongst the stars. The names of meteor streams such as Taurids and Orionids refer to the positions in the constellations of Taurus and Orion from which the meteors would be seen to radiate. Could it be that the ancient Egyptian preoccupation with Orion was due to the fact that a point in Orion's belt appeared to be the source of meteors and deadly missiles that threatened their empire over many decades? Or that the channel in the Giza pyramid, that pointed to the star in the constellation of Draconis, the dragon, really drew attention to the apparent source of cosmic dragons that plagued their lives?

The enormous stresses that were caused by unrelenting assaults from the sky could well have strengthened the resolve of Egyptian Pharoahs to combat adversity through feats of technological achievement. The construction of the Giza pyramids must certainly have represented a huge advance on the prevailing technologies of the day. The average weight of each stone block that makes up a pyramid is over two tons, and it seems certain that such blocks had to be transported very great distances across the desert. This was surely an enterprise that has no parallel in the entire history of human civilisation. How exactly the pyramids of Giza were constructed remains an enigma even to the present day. On the scale of difficulty it would make the construction of a great cathedral pale into utter insignificance.

Why in the first place did the Egyptians choose to build such stupendous structures if they served no function except as royal tombs? Why did they scatter them over a vast area of desert, rather than build them all in one place? One thing is certain. The pyramids were planned and built with meticulous car and their construction involved an effort that comes close to being superhuman. The Giza pyramids have survived for 4,500 years, surely withstanding several episodes of assaults from the skies. Hoyle suggested some years ago that a solid pyramid may have just the right structure to maximise the chances of surviving the blast wave from Tunguska-type explosions in the sky. One could not devise a better monument to survive millennia rather than decades or centuries.

My own favourite theory is that the pyramids were not just royal tombs, nor were they simply religious icons, but they actually served the kings before their death, as a kind of air-raid shelter, to protect them from the explosions of cosmic missiles. The channel within the Great Pyramid pointing in the direction of Orion's belt could, with the deployment of a single plane mirror or polished surface, have been used to observe the progress of an offending meteor stream, focusing on the point on the sky from which the missiles would have appeared to come. Other channels through the pyramid could have been used for other purposes: life support perhaps, ventilation, and transport of food.

Archaeological discoveries by Donald B. Redford of Penn State University point to the remains of what may have been cosmic air-raid shelters for lesser mortals near the end of the Old Kingdom. Groups of bodies have been discovered with arms placed over their heads and bodies in various contorted positions, exactly as they fell, strongly suggesting that they were the hapless victims of an unexpected assault from the skies. The bodies are found to be associated with remnants of curved walls that look very much like parts of an air-raid shelter that did not survive.

More compelling evidence for a cosmic cataclysm ending the Old Kingdom comes from work of scientists at Queen's University Belfast led by Mike Baillie. Baillie is one of the pioneers of the new science of dendrochronology which involves the study of the thicknesses of tree rings at different times in the past. Each year a new ring is formed in the trunks of trees, resulting from growth during the summer months. A narrow tree ring corresponding to a particular year means that there was little tree growth during that year, which could only be due to greatly diminished levels of sunlight. Such a thinning of tree rings in Irish oaks has actually been discovered over the entire period 2354–2345BC, which comes close to the final decades of the Old Kingdom. This is most easily explained as being due to the arrival of Tunguska-type cometary missiles that would have dusted the atmosphere and dimmed the light from the sun.

The next well documented episode of bunched collisions may have occurred over a millennium later during the period 1350–1100BC. Here again Baillie's evidence from tree ring thickness measurements show a marked climatic downturn at 1159–1141BC which could have been caused directly by cometary missile impacts or by volcanoes that were triggered by such impacts. At the start of this period we find some suggestive evidence from the desert sands in Arabia. Whenever desert sand became violently shocked by explosive events the sand would turn into a glass that possesses very special reflective properties. Such glass could be used as precious ornamental gem stones that are set in items of jewellery. The priceless treasures of Tutankhamen have now been found to include a necklace that contains just such an impact-shocked gem stone as its centrepiece. The necklace adorns none other than the boy emperor Tut who reigned from 1361 to 1352BC, although the stone itself could have been collected during earlier impact epoch. An elaborate motif on the necklace contains symbols of Sun and

Moon, suggesting that celestial provenance of the central stone was recognised.

The dates of events described in the Old Testament are open to dispute, but some at least may have occurred between 1300 and 1100BC. Many of the Old Testament accounts of seemingly bizarre and mysterious occurrences could have had a firm basis in fact if one admits the possibility of an epoch of bunched cometary impacts. Descriptions of deluge, a rain of fire on the cities of Sodom and Gomorrah, famines occasioned by the wrath of the gods – all have a rational basis as possible effects of cometary impacts. Fires, tsunamis or tidal waves, floods, climatic changes adverse to crops, even clusters of earthquakes, can now be interpreted as real phenomena caused by the arrival of cometary missiles. No metaphysical or mystical explanations are required. We can also begin to comprehend what it was that Joshua saw when he reported that the Sun stood still in the sky. It was almost certainly the glow of an immense fireball similar to what was seen over Tunguska in June 1908. The two sets of descriptions, in the Old Testament and in Siberia of 1908, are almost identical.

At roughly the same time in China, close to 1100BC, there were reports of 'widespread disturbances of the barbarians' according to bronze inscriptions of the Western Chou (c. 1122–771BC) and this nearly coincided with the fall of Jericho and the beginnings of Judaism in the West. During this period we are also told that the authority of the Chinese emperors became reinforced after perhaps diminishing for a short while. A link was then forged between the authority of the king and the power of Heaven over the Universe, and subsequently kings were regarded as the Sons of Heaven. It is possible that the use of the dragon motif that came to be used extensively as this time was occasioned by an episode of 'dragon' invasions from the sky.

We have already noted that the break-up of comets during collision episodes would have frequently led to spectacular displays in the ancient skies, and that these in turn would have given birth to celestial combat myths in ancient societies. Such celestial combat myths involving wars of the gods are clearly evident in Greek traditions as in the Homeric poems around the eighth century BC. It is likely that Greek mythology evolved from the older mythologies of Western Asia and Mesopotamia that date back somewhat before 1100BC when the Earth had suffered an episode of severe bombardment by cometary bolides.

Even some 200 years after the Homeric poems were composed, the Greece of classical times in the fifth century BC had not entirely forgotten the bolts from the blue, the disastrous events of bygone centuries. In the dialogues of Plato we thus find references such as the following in *Timaeus*:

Critias: . . . There were of old great and marvellous actions of the Athenians, which have passed into oblivion through time and the destruction of the human race . . . there have been, and will be again, many destructions of mankind arising out of many causes; the greatest have been brought by the agencies of fire and water . . . There is a story . . . that once upon a time Phaethon, the son of Helios, having yoked the steeds in his father's chariot, because he was not able to drive them in the path of his father, burnt up all that was upon the earth, and was himself destroyed by a thunderbolt. Now, this has the form of a myth, but really signifies a declination of the bodies moving around the earth and in the heavens, and a great conflagration of things upon the earth recurring at long intervals of time; when this happens, those who live upon the mountains and in dry and lofty places are more liable to destructions than those who dwell by rivers or on the seashore. When, on the other hand, the gods purge the earth with a deluge of water, among you, herdsmen and those of you who live in cities are carried by the rivers into the sea.

Dialogues of Plato (Timaeus), translated by B. Jowett

It is clear from this that stories of comet and meteorite impacts "recurring at long intervals of time" were in currency right through to the time of Plato. The intellectual movement to forget and bury a turbulent past, and to deem the Earth safe from comets, seems to have begun in earnest with the philosopher Aristotle who lived in the period 384–322BC. With Aristotle, comets and meteors lost their exalted status as celestial objects, and were essentially downgraded to become relatively minor effects in the atmosphere. The Earth became disconnected from the cosmos, at least in Western thought. It is possible that such changes of attitude were made possible, scarcely 200 years after Socrates, for the simple reason that there then was a marked decline of all forms of meteor and cometary activity.

A mainly benign period appears to have started about 1000BC and continued, with a few remissions, through classical times until the sixth century AD. Events in the sixth century bear all the hallmarks of another episode of cometary missile impacts, although perhaps less intense than at earlier times. The collapse of the Roman Empire in the sixth century AD has been the subject of intense scholarly debate for many years. Edward Gibbon's account of what happened leaves little room to doubt the reality of geological disturbances that played an important part in the final stages in the disruption of the Empire. The description goes thus:

> . . . history will distinguish . . . periods in which calamitous events have been rare or frequent and will observe that this fever of the Earth raged with uncommon violence during the reign of Justinian (AD527–65). Each year is marked by the repetition of earthquakes, of such duration that Constantinople has been shaken above forty days; of such an extent that the shock has been communicated to the whole surface of the globe, or at least of the Roman Empire. An impulse of vibratory motion was felt, enormous chasms were opened up, huge and heavy bodies were discharged into the air, the sea alternately advanced and retreated

beyond its ordinary bounds, and a mountain was torn from Libanus and cast into the waves . . . Two hundred and fifty thousand people are said to have perished . . . at Antioch.

Edward Gibbon, *Decline and Fall of the Roman Empire*

The type of prolonged and frequent earthquake activity of the type described here is of course unusual. As we have stated before in relation to volcanic eruptions, the explanation has to be sought in terms of an external cause – cosmic missile impacts that could provide the trigger to send pressure waves into the Earth's crust and thereby generate prolonged bursts of seismic activity.

Evidence of another major downturn in the Earth's climate has again come from work by Baillie. His study of tree ring thicknesses in Irish oaks shows that there was little tree growth during the years around AD540. Similar studies by others have also shown the same effect – narrow tree rings at this time – in places as wide afield as Germany, Scandinavia, Siberia, North America and China. The idea that a volcanic eruption was responsible for a dust shroud that lowered the temperature and reduced seasonal tree growth does not tally with the lack of an acid signal in Greenland ice-drills of this same period. Furthermore, volcanic dust is known to settle in a couple of years at most, and cannot therefore explain such a protracted episode (AD535–46). There can now be little doubt that a major global catastrophe enveloped the planet around the year AD540.

The time of this climatic downturn – as already noted, coincides with the final stages of the collapse of the Roman Empire. The year AD540 also marked the beginning of the 'Dark Ages' in Europe. Mike Baillie has cited an interesting remark attributed to Roger of Wendover written in 540 or 541 referring to 'a comet in Gaul so vast that the whole sky seemed on fire. In the same year there dropped real blood from the clouds . . . and a dreadful mortality ensued'. The death of King Arthur at around 540 is also claimed by Baillie to have had possible links with myths of fire from the skies and the concept of the 'waste land' that devastated Britain in the middle of the sixth century AD: once again tying in with the idea of bunched cometary impacts.

There is also a convergence of imagery from Celtic mythology of the same time. A famous legend featuring Merlin appears to be based on an historical event, the Battle of Arfderydd of AD540. The story of Merlin (*Myrddin* in Welsh) has versions in Wales, Ireland and Scotland, and it is the Scottish version (where *Myrddin* is called *Lailochen*) that has the most explicit imagery relevant to our story. According to a translation by Basil Clarke, Lailochen makes the following pronouncement as he flees into the woods from the battlefield:

> In that fight the sky began to split above me, and I heard a tremendous din, a voice from the sky saying to me 'Lailochen, Lailochen, because you alone are responsible for the blood of all these dead men, you alone will bear the punishment for the misdeeds of all . . . But when I directed my gaze towards the voice I heard, I saw a brightness too great for human senses to endure. I saw, too, numberless martial battalions in the heaven like flashing lightening, holding in their hands fiery lances and glittering spears which they shook most fiercely at me.

Other evidence from contemporary literature comments that 'the sun was dark and its darkness lasted 18 months . . . the sun appears to have lost its wonted light and appears of a bluish colour . . . fruits did not ripen . . . cold and drought finally succeeded in killing off the crops in Italy and Mesopotamia and led to terrible famine in the immediately following years'.

Intense meteor and cometary activity has been recorded in the closing decade of the sixth century AD. For instance in an Islamic text one finds the following graphic description:

> In the year 599 on the last day of Moharrem, stars shot hither and thither and flew against each other like a swarm of locusts;

people were thrown into consternation and made supplication to the Most High.

In India it appears that the dawn of the sixth century AD marked a time of exceptional turbulence and strife. In *A History of India* J.C. Powell-Price writes:

> The eruption of the Huns had a very great effect on India. It broke up the empire of the Guptas and upset the developments which the Gupta culture was leading to. These barbarous hordes were out for plunder, blood and rape and had no feelings for either art, literature, or religion . . . The history of N. India is very confused after the eruption of Huns in about AD500.

The cause of such disturbances is not made clear, but one could not imagine a better backdrop for anarchy and chaos than cometary impacts. Such events would undoubtedly have shaken the very foundations of society and weakened the capacity of authorities to enforce law and order.

Oriental religions echo the hypothesis of cometary impacts. Hinduism could be seen as a religion of forbidding austerity, with its formidable pantheon of gods and goddeses, all dwelling in the skies. Even though the many deities of Hinduism are often considered as manifestations of Brahma, they nevertheless have individual identities and lead separate lives. Vedic Hindu traditions apparently began with the migration into India of Aryan-speaking tribes from further west. Hinduism may have begun around 2400BC, possibly during a bad period of cometary missile impact, and so, understandably, there is a firm acknowledgement of Man's connection with the external Universe. Hindiusm saw the world as an uneasy balance between order and the powers of chaos. A mythological expression of this philosophy is seen as an ongoing conflict between gods (devas) and disruptive antigods, 'asuras'. This is reminiscent of other combat myths that could be identified with a perceived battle of dragons in the sky. The Hindu concept of Trimurti is also of possible relevance in this context. The Trimurti consists of Brahma, the Creator, Vishnu, the redeemer, and Shiva, the destroyer. The world is thus considered to go through repeated cycles of creation, destruction and redemption, which is just what the cosmic dragons would do.

Ancient Chinese religions, as far as we know, all acknowledged the power of Heaven (the skies) over men. Kings derived their authority essentially to be a divine act of delegation of power. In almost 4,000 years of civilisation China would have expeienced a few 'bad' periods of bunched impacts which it successfully weathered. On the whole the ancient Chinese were governed by an overriding pragmatism and a stoical acceptance of natural events. Chinese astronomers were meticulously accurate observers of the skies and never once allowed prejudice or fear to hinder their work. Nature was there to be observed and studied, not to be tinkered with.

The holistic philosophies of the South Asia and China must have gone a long way towards helping the people of this continent cope with the less desirable manifestations of the cosmic dragons – fires, floods, destruction and death. The good times and the bad times had to be considered together, making up a natural balance of the Universe as a whole. During the calm periods that were free of impacts no amnesic barriers were erected to disconnect the Universe from Man. In this regard it should be stressed that attitudes differed markedly from those in the West. The Universe and Man remained intimately and inextricably linked.

References

BAILLIE, M.G.L. (1994). Dendrochronology raises questions about the nature of the AD536 dust-veil event, *The Holocene*, 4, 212–217.

BAILLIE, M.G.L. (1996). Chronology of the Bronze Age: 2534BC–431BC, *Acta Archaeologica*, 67, 291–298.

BAILLIE, M.G.L. (1998). Exodus to Arthur: close encounters with comets and the fiery dragons of myth, London: Batsford.

BAUVAL, R. & GILBERT, A. (1994). *The Orion Mystery*, London: Heinemann.

BRÜCK, M.T. (1995). Can the Great Pyramid be astronomically dated? *J. British Astron. Assoc.* 105, 161–164.

CLUBE, V. & NAPIER, B. (1990). *The Cosmic Winter*, Oxford: Basil Blackwell.

CLUBE, V. & NAPIER, B. (1982). *The Cosmic Serpent*, New York: Universe Books.

CLUBE, S.V.M., HOYLE, F., NAPIER, W.M. & WICKRAMA-SINGHE, N.C. (1996). Giant comets, evolution and civilization, *Astrophysics and Space Science*, 245, 43–60.

GUNN, JOEL D. (ed.) (2000). The years without summer: Tracing AD536 and its aftermath, BAR International Series 872.

HOYLE, F. (1993). *The Origin of the Universe and the Origin of Religion*, Rhode Island: Moyer Bell.

PEISER, BENNY J., PALMER, TREVOR & BAILEY, MARK E. (eds) (1998). Natural Catastrophes During Bronze Age Civilization, BAR International Series 728.

TRIMBLE, V. (1992). 'Cheops' Pyramid' in *Visit to a Small Universe*, New York: American Institute of Physics.

9. Asteroid Threat in
Modern Times

At the closing of each Kalpa, dear Prince
All things which be will back to My Being come;
At the beginning of each Kalpa, all
Issue newborn from Me.

Bhagavad Gita

In the previous chapter we examined the possible effects of the Earth interacting at more or less regular intervals with the fragments of a giant comet that remained in Earth-crossing orbits. The operation of such a scheme of events may require special conditions of approach past Jupiter and Saturn, and special relationships beween the orbital periods of the comet and those of Jupiter and Saturn, but these conditions cannot be ruled out as impossible. We approached the problem not from a rigorous, dynamical standpoint but from a heuristic point of view, seeing what the events of history told us about major impact episodes. These events did indeed seem to indicate the recurrence of bunched collision epochs, starting with a series of collisions that led to the emergence of the Earth out of the last ice age. The latest in a series of such impact episodes was seen to have occurred at about 1300BC and AD540, the last of which was probably far less intense than earlier ones. With the original meteor stream becoming progressively denuded and more dilute as time goes on, the next in this series of encounters when it comes is likely to be less severe.

Whilst the interaction of the Earth with a single giant comet and its fragments could be supposed to occur cyclically with high risk conditions being concentrated over several decades, the average risk at other times is likely to be dominated by another class of object – the asteroids. There are a vast number of asteroids composed mostly of rock and rubble measuring from a few hundred kilometres to less than a kilometre across, orbiting the Sun between the orbits of Mars and Jupiter. Astronomers had for a long time expected a planet to lie between the orbits of Mars and Jupiter and were puzzled by its absence. The total amount of material in the form of asteroids is comparable to the masses of the terrestrial planets, which led to an early theory that these objects were in fact the result of disintegration of an Earth-like planet. That theory has now been superceded by another, which as we discussed in Chapter 2, asserts asteroids to be the bits of planet-type material that failed to accumulate and grow into the size of a full-blown planet. Such a failure might have been due in part to the disruptive action of Jupiter's gravity that would have constantly stirred up the asteroids, thus preventing them from coming together into a single planetary body.

The first asteroid was discovered in 1801 by the Italian astronomer Giuseppi Piazzi and was named 1 Ceres. Since then asteroids have been discovered at an ever increasing rate. Some 20,000 or so asteroids have been discovered to date and a record of their orbits and details are kept in a 'minor planets' database currently maintained at Harvard University. Asteroids are designated by a number and a name – e.g. 2 Pallas, 433 Eros, 2001 Einstein, 8077 Hoyle.

The vast majority of asteroids for which orbits have been studied are found to pursue paths that do not bring them anywhere within reach of the Earth. The first Earth-approaching asteroid 433 Eros was discovered as far back as 1898. This 34-kilometre long sausage-shaped asteroid was the target of the NASA spaceprobe NEAR (Near Earth Asteroid Rendezvous) Shoemaker which was launched in 1996. The probe orbited the asteroid for several months before crashing on to its surface on 12 February 2001. The spacecraft transmitted dramatic views showing a boulder-strewn and mildly cratered surface, and as it crashed even more detailed pictures were received (Plate 9.1). The evidence accumulated in the seconds before the kamikaze crash is expected to yield important information about the detailed structure of the asteroid and vital clues as to how asteroids may have been formed.

Over the past two decades many asteroids that are on a possible collision course with the Earth have been identified. Systematic examinations of old astronomical plates combined with new tele-scopic searches for faint objects have led to the discovery of over 1,000 small asteroids in orbits that intersect the orbit of the Earth. These orbits evolve in a chaotic way because of the unpredictability of the perturbing influences of planets many orbital periods into the future. Long-term forecasting of collision probabilities are therefore very difficult to make. Numerical modelling of asteroid orbits using fast computers, in which orbits are followed forward in time is being carried out by Duncan Steel of the University of Salford. One could hope that Steel's calculations would eventually narrow down the list of asteroids that pose a serious threat to our planet.

In Chapter 8 we attributed the Tunguska event of 1908 to a fragment of a comet that exploded in the atmosphere. There has, however, been a long-standing argument as to whether the object in question was a cometary fragment or a small asteroid. On balance a cometary object is more likely, mainly for the reason that no material of meteoritic origin was recovered from the Tunguska site.

Plate 9.1 (opposite page)   The asteroid Eros. Top frame shows a distant view; Bottom frame a close up of the surface photographed in February/March 2001. (Courtesy © NASA)

If the Tunguska event was caused by a rocky asteroid the estimate of its dimension would be slightly smaller than that for a cometary bolide – about 50 metres across, and the energy released equivalant to about 15 megatonnes of TNT. Such an asteroid collision if it happened today over central London would flatten the city out to the M25 motorway and beyond and lead to millions of deaths.

It is difficult to estimate in a reliable way the chances of a Tunguska-type asteroid event happening at the present time, say within the next year or within the next decade. The reason is that only a very small fraction of the potentially hazardous objects in our neighbourhood have been charted to date and the list is always going to be incomplete. So the fateful event when it does happen will come literally as a bolt from the blue. There is also the remote chance of a rogue long-period comet stealing upon us with no warning whatsoever, approaching along an Earth-crossing orbit and striking us unawares.

The scale of the damage to be expected from such a collision, whether an asteroid or comet, depends mostly on the size of the object. A comet is in general a little more dangerous than an asteroid of the same size. Comets have impact speeds somewhat greater than those of asteroids on average, but with the bulk density of an icy comet being about half that of an asteroid made of rocky material, the margin of difference is minimised. The effects to be expected from the impact of comets of different sizes is set out in Table 9.1.

The possibility of a full-blown 10-kilometre comet collision happening in any particular year is perhaps of the order of one in 100 million. Such a collision will produce effects similar to those that led to the extinction of the dinosaurs 65 million years ago, and is described in the last row of Table 9.1. A large fraction of the world's population will be killed. The probability of anyone dying from such a collision in the next year is therefore one in, say, 200 million, compared to one in 14 million chance of one winning the UK national lottery.

As far as a comet impact is concerned all one could do is to take such a risk in one's stride. The more immediate risks that might be averted arise from impacts much higher up in Table 9.1, asteroid or cometary bolides of the Tunguska type. It would seem our bounden duty as citizens of the twenty-first century high-tech world to do whatever is possible to track down potentially hazardous asteroids and hopefully to avert any impending disaster. Many nations, including Britain, are currently undertaking research programmes designed to increase our knowledge of asteroids and consequently to improve our assessment of the immediate threat posed by such objects.

Since the public has become aware of the comet and asteroid threat several near-misses have come to light. On 15 March 1994 a

Table 9.1 Effects of asteroid impacts

Energy	Diameter for ice comet or comet fragment with v=13.5km/s	Effects
1 MT TNT (50 Hiroshima bombs)	70m	Disintegration at high altitude, h>30km
10MT TNT (500 Hiroshima bombs)	140m	Explosion in low atmosphere or surface impact; localised blast
1000MT TNT (50,000 Hiroshima bombs)	700m	Extensive damage; fires; tsunamis
100,000MT TNT (5 million Hiroshima bombs)	3.2km	Extended cratering; global effects; climatic catastrophe; crop failures; famines
1,000,000MT TNT	7km	Mass extinctions

very small asteroid measuring a dimunitive 10 metres across passed by the Earth at less than half the distance between the Earth and the Moon. Similar near-misses are estimated to take place as frequently as once a week. They go unnoticed because there are no systematic surveys to monitor the skies for such small objects that become visible only when they come very close to the Earth.

Ignorance of the precise asteroidal traffic flow past the Earth has triggered a few Armageddon-style scares in recent years. In late 1992 the big story that hit the news was that a comet called Swift-Tuttle would crash with the Earth in the year 2126. The comet, named after two astronomers Lewis Swift and Horace Tuttle who first spotted the object in 1862, is the source of the Perseid meteor stream. The Earth crosses this stream every August giving rise to the well-known Perseid meteor showers. The earliest observations of the parent comet of this shower, Comet Swift-Tuttle, yielded an estimate of its orbital period as 120 years, which turned out to be inaccurate. When the comet failed to return after 120 years in 1982 Brian Marsden urged astronomers, both professional and amateurs, to search for the return of the comet, particularly following a spectacular Perseid shower which was seen in August 1991. The comet was duly discovered and found to have reached its last perihelion (closest approach to the Sun) in 1993; and Marsden's computation of the comet's orbit now predicted a return in 2126. The comet's orbit with a present-day period of 131 years is found to be 'locked' to Jupiter in a way that it would continue to make close approaches to the Earth periodically for the next 10 millennia. But the comforting news for the moment is that the comet would not collide with the Earth in 2126.

After Comet Swift-Tuttle several other false alarms of collisions or near-misses have been raised. In 1998 asteroid watches provoked an international scare by claiming that the 2 kilometre-long asteroid XF-11 will hit the Earth in 2028. A few days later the threat was withdrawn after more detailed calculations of its orbit showed that it would pass by the Earth at a relatively safe distance. In 2000 another scare on a much smaller scale had surfaced. An asteroid called 2000 SG344, some 30–70 metres across, appears to be destined to collide or come very close to colliding with the Earth on 21 September 2030. The risk of the collision is now estimated at about one in 500, but nearer the day this esimate could be fine-tuned, either to increase or reduce the odds of a collision. The impact of SG344 on the Earth would release an amount of energy corresponding to 50–100 equivlaent Hiroshima bombs. What society would elect to do in the immediate run-up to this event is of course left to be seen. Evacuating a precisely calculated target area may in such a case be more sensible than making any great effort to deflect the incoming missile.

The year AD2030 is not too far way, and there is no guarantee whatsoever that SG344 will be the first Tunguska-type asteroid to come our way. An asteroid or a cometary bolide of diameter well below 50 metres would explode high in the atmosphere and would not lead to much damage. A bolide significantly larger, similar to Tunguska, will explode near enough to the ground for its shock waves to cause considerable local damage. US military surveillance data that was declassified only in 1993 revealed that between 1975 and 1992 some 136 explosions were detected in the atmosphere by space-based infrared and visual detectors. These were caused by near-Earth objects (NEOs as they are now called) of diameters well below the critical 50 metre limit.

Present-day estimates of the number of potentially hazardous NEOs are still woefully inadequate. Although an average frequency of one in a few hundred years is still widely quoted for Tunguska events in the present day, the huge uncertainities in the total number of 100-metre sized objects in our vicinity could render this estimate wildly inaccurate. The frequency could well be once in 30 years or even more frequent, in which case it would be prudent to have contingency plans well in place.

The first step in such a plan of action would be to enlarge the NEO database with telescopic searches as were pioneered some years ago by Tom Geherels of the Univeristy of Arizona using the 0.9 metre telescope on Kitt Peak in his Spacewatch effort. This survey alone found the number of NEOs with diameters less than 50 metres crossing the Earth's orbit to be a hundred times more was thought previously. The original Arizona effort is now being augmented by other programmes, including the Palomar Asteroid and Comet Survey (PACS), which uses the 0.46 metre Schmidt telescope on Mt Palomar, and the Anglo-Australian Near-Earth Asteroid Survey (AANEAS), and there are others in line to join them. Such a global collective effort has been dubbed 'Spaceguard' after the name of a fictional spacewatching service described by Arthur C. Clarke in his novel *Rendezvous with Rama*.

But what then if Spaceguard or another similar survey pointed to a large object hurtling towards us on a collision course? What does one do? A great deal would depend on how much advance warning we get of such an impending event, and on how big the object turns out to be. If it is small like SG344 then to evacuate the danger zone and keep well out of the way might be the best. If it is a much larger asteroid, well over a 100 metres in diameter, and it is detected several orbital periods before the predicted collision event, the first step could be to send out a space probe to reach the object in order to determine its precise size and structure and hence to assess the exact nature of the threat. For a small and relatively loose aggregate of asteroidal matter one might contemplate dispersing it with a moderate explosive power into small enough fragments that were each below the critical size for ground impact. Indeed it may be possible to turn the hazard into a spectacular fireworks show that would end benignly in the high stratosphere. A large compact asteroidal mass could pose more serious problems, however. Strategies currently under discussion include nudging the object sideways to deflect it into a safe orbit with the use of nuclear explosions or the nuclear-powered rocket thrusters. All this is still in the realms of distant planning, but progress is slowly being made.

References

CLARKE, A.C. (1990). *Rendezvous with Rama*, London: HarperCollins.

GEHRELS, T. (ed.) (1994). *Hazards due to Comets and Asteroids*, University of Arizona Press.

STEEL, D. (1997). *Rogue Asteroid and Doomsday Comets* Wiley.

STEEL, D. *et al.* (1997). Anglo-Austrlian Near-Earth Asteroid Survey: Valedictory Report, *Austr.J.Astro.*, 7, 2, 67.

10. Dragons Breathe Death

By some physicians influenza was supposed to be
contagious; by others not so; indeed its wide and rapid
spread made many suspect some more generally
prevailing cause in the atmosphere . . .
Robert Thomas, *The Modern Practice of Physic, 1813*

Our ancestors of a distant past were apparently unanimous in the belief that comets were the cause of disease and pestilence. The association between comets and epidemic disease appears to have crossed the boundaries of geography and culture. We tend nowadays to dismiss these ancient ideas as primitive superstition born of ignorance. But was this really so? Our forebears were probably more objective and pragmatic in the way they looked at the world. They were not so rigidly bound to dogma as we are. Nor were they constrained by the authority of institutions that deemed what was respectable and proper to believe and what was not. So any ancient assertion that one natural phenomenon A is linked to another B should perhaps be taken at face value, at any rate as a working hypothesis worthy of further exploration. The implication would be that a correlation between phenomenon A and phenomenon B was verified over many generations and this information was transmitted through oral tradition. In general it would follow then that the older a tradition the greater is the likelihood of a factual basis and the more respect it therefore deserves.

136

Following an epoch of bunched collisions by cometary missiles it would be natural to expect large-scale devastation of the land, crop failures and consequently famines. One might imagine that epidemics of disease followed quite naturally from the extended period of malnutrition of a population that may have lasted for many years. This might of course account for some epidemic diseases, but not all. Historical epidemics such as the Plague of Athens, the Plague of Justinian and the Bubonic Plague would hardly qualify to be classed as diseases of poor nutrition.

Direct historical evidence linking particular conspicuous events in the sky to epidemics is hard to come by. An exception could be the record we already mentioned of 'a comet in Gaul so vast that the whole sky seemed on fire . . .', with the great Plague of Justinian immediately following in the year AD542.

The connection between comets and epidemic disease cannot be properly understood independently of the wider issues relating life to comets. We argued in earlier chapters that life could not have originated *de novo* on the Earth, and the picture that emerged was that interstellar space is filled with living cells, cells which are mainly in a frozen dormant state. Every niche suitable for life that develops in the cosmos, through collapse of cosmic gas clouds and the formation of stars, comets and planets becomes quickly infected with this all-pervasive living system. Such 'infection' leads to a vast increase in the numbers of bacteria, with an inevitable feed-back of biomaterial into the space between stars. It then becomes *almost* meaningless to pose the question: in which star system or in which corner of the Universe did life begin? Life is a truly cosmic phenomenon, the sum total of an evolutionary experience gained and accumulated in a multitude of different places, quite widely separated, and extending in time throughout the entire history of the Universe.

In our own solar system, we argued that life in the form of bacterial cells were first housed in the comets. Every one of nearly 100 billion comets had, at the beginning, warm watery interiors and

137

all the organic nutrients necessary to make for a congenial breeding ground or microbial life. We mentioned in earlier chapters that comets crashing on to the Earth brought the oceans and the atmosphere. With the oceans and the atmosphere so deposited, comets would also have seeded our planet with life – a life that under the protected canopy of cloud-covered skies was able to take root and flourish.

The first successful seeding of the Earth with life occurred at about 4 billion years ago. But according to this picture the process of cometary seeding could not have stopped at a distant geologic time. The Earth is well and truly entwined in the débris that is shed from comets. About 100 metric tons of cometary débris, much of it organic and biological, enters the Earth's atmosphere every day. A lot of this débris is either sterile, due to exposure to the Sun, or is burnt up on entering the Earth's atmosphere. But it is inevitable that a small fraction of the incoming dust, freshly evaporated from comets, must contain active or viable microbes that actually survive entry through the Earth's atmosphere. The total number of bacteria thus arriving per year could be as high as 10^{22}. And the total number of viruses could number upwards of 10^{23} per year, exceeding the number exuded by the entire human population of the Earth by several powers of ten.

We have so far not referred to viruses in our development of the theory of cosmic life. The first clue as to the nature of viruses came as early as 1898 when Friedrich Loeffler and Paul Frosch discovered that the causative agent for foot-and-mouth disease in livestock was a particle that was smaller than any known bacterium. Viruses occupy a grey area in biology between living and nonliving states. Their essential components involve a protein coat or capsid enveloping a genome comprising either DNA or RNA that codes for function. A virus can replicate only within the cell of the host it infects, and such infections are known to cause a large number of diseases in plants and animals. Examples of human diseases caused by viruses include the common cold, influenza, smallpox, polio and AIDS.

The behaviour of a virus as it infects a host cell cannot be better described than the account given in Sir Christopher Andrewe's book *The Common Cold* published in 1965 by Weidenfeld and Nicholson:

What happens when a virus infects a cell is probably something like this. The protein part of a virus makes specific contact with something on the cell-surface. Then the cell ingests or takes the virus up within itself. It may ingest the whole virus and break it up inside or, as happens with the bacteriophages or viruses infecting bacteria, the protein coat of the virus may be left outside the cell, only the essential nucleic acid gaining access to the interior. In either event, the protein part of the virus is expendable and plays no further part. The nucleic acid part, however, proceeds to instruct the cellular mechanism in a sinister manner. Suppose it is a rhinovirus infecting a cell lining your nose. The instruction will run thus: 'Stop making the ingredients necessary for making more nose-cells. Henceforth use your chemical laboratory facilities for making more nucleic acid like me.' The intruding virus nucleic acid gives the further instruction: 'And now make a lot of protein of such-and-such composition which I require wherewith to coat myself'. The cell can do nothing but obey and as more new virus particles are thus assembled by the cell's chemical machinery they are, at the end of the production line, turned out into the outside of the cell. With many viruses, including probably rhinoviruses, the final effect is to exhaust the cell altogether, so that after a while it dies and disintegrates. The virus set free will infect more of its victim's cells until such time as defence mechanisms have been mobilised. It will also get into the outside world and infect more victims, for one result of the cell-destruction in the course of a

cold infection will be inflammation, pouring out of fluid, sneez-
ing and spread of virus. All of it a very conveniently organised
affair for the benefit of the virus . . .

Viruses are normally known to be able to swap genetic information between species, and this property is used nowadays for
medical applications and in genetic engineering. Viruses could also
add naturally to the genomes of creatures they infect. It is of
interest in this context that eukaryotes, including human cells, have
substantial stretches of viral DNA that is normally unexpressed.
The silent viral DNA could play a part in the evolution of species,
and it could be speculated that without this viral component the
process of evolution itself might be severely curtailed.

Viruses are normally associated only with eukaryotic cells of
higher life-forms. The corresponding infective agents of bacteria
are called plasmids, and these could also exchange genetic information between species. The interaction of a plasmid with a bacterium
could, for instance, make a nonpathogenic variety into a pathogenic
form.

Until recently it was thought that the total number of distinct
bacterial species on the Earth was a mere few thousand. Today,
using modern techniques to detect sequences of nucleic acids rather
than intact bacterial cells, it is estimated that the total number of
bacterial species might be well in excess of a billion. The vast
majority of these bacteria are thought to be what are called
extremophiles, bacteria seeking extreme and hitherto uncharted
environments. They have not yet been cultured, perhaps never will.
They lie in surface soil and surface water, evidently doing nothing
– perhaps they are waiting for the right host to emerge. Perhaps
they are falling from the skies.

A direct way of testing this latter possibility exists if this
microbial input includes micro-organisms that are pathogenic to
plants and animals. Rather as physicists use amplifying electronic
counters to detect very small fluxes of incoming cosmic ray

particles, so in a similar way plants and animals could be regarded as amplifying detectors for microbes from space.

In general, an incoming pathogen (virus or bacterium) from space has one of two logical paths to follow. It can either establish a reservoir in some group of plants or animals, and thereafter proceed to propagate by horizontal transmission; or it could have the property of not being able to form a stable reservoir. In the first case, of which smallpox and AIDS are examples, epidemics would arise whenever a reservoir effectively breaks its banks. In the second case, every epidemic, every single attack, must be driven directly from space. We shall see later in this chapter that this may be the case for influenza.

A reading of medical history provides ample evidence for such invasions from space and of a changing pattern of human diseases. Many bacterial and viral diseases have a record of abrupt entrances, exits and re-entrances on to our planet, exactly as though the Earth was being seeded at periodic intervals. In the case of smallpox, a disease caused by a virus, the time interval between successive entrances seems to have been about 700–800 years. Since man is the only host of the smallpox virus, global remissions of this disease lasting for many hundreds of years are very hard – almost impossible – to understand. From the conventional point of view one has to say that the virus became extinct, and then re-evolved to precisely its original form from some unknown ancestor after many hundreds of years. A rather improbable event that must be. From skin lesions discovered in Egyptian mummies one could argue a case for smallpox being prevalent a few hundred years before the classical period in Greece. More decisively, perhaps, we could argue that it was present a few hundred years after the dawn of the Christian era. With the accurate medical records that are available at the time we can be equally sure that smallpox was *not* present in classical Greece and Rome. In fact the Latin word for pock mark which is *variola* is said to have been coined only in seventh century

AD, again confirming that smallpox was absent in the early Christian era. A genuine world-wide absence would seem to be implied at this time, for otherwise it would be hard to imagine a disease so infectious as smallpox being kept out for so long from the Western world.

A bacterial disease that followed a similar pattern of recurrent entrances and exits over the centuries is the bubonic plague. The first thing to note is here that the bubonic plague is not primarily a disease of man. The bacterium *Pasteurella pestis* which causes the plague attacks many species of rodent, as its first target. Because the black rat happened to live in close proximity to humans, nesting in the walls of houses, the physical separation of people from the black rat was always small and could be bridged by fleas. Fleas carried the bacterium from the blood of the rats to the blood of humans.

Although the transfer of the bacterium from human to flea and back to human presumably occurred, it does not seem to have been sufficient to maintain the disease. Every epidemic of the disease died out as the supply of rats became exhausted. As with smallpox, bubonic plague has come in sudden bursts separated by many centuries, and there are the same difficulties of understanding where *Pasteurella pestis* went into hiding during the long intermissions. A somewhat ambiguous reference occurs in the Old Testament, at a date of about 1200 BC, when the Philistines are said to have been attacked by emrods [buboes] in their secret parts . . .' as a reprisal of God for an attack on the Hebrews. A clear reference to bubonic plague also occurs in an Indian medical treatise *Charaka Samhita* written in the fifth century BC, in which people are advised to leave houses and other buildings, 'when rats fall from the roofs above, jump about and die'.

There may have been an outbreak of plague during the first century AD, with centres of the disease in Syria and North Africa, but between the first and sixth centuries there were no known attacks of the disease. In 540 AD, a pandemic involving the Near

East, North Africa and Southern Europe is said to have had a death toll that reached 100 million, with more than 5,000 dying each day in Constantinople alone. We have already noted in an earlier chapter that this coincided with a bunched cometary collision episode that was marked by a global climatic downturn. The epidemic was the so-called Plague of Justinian, Justinian being the Roman Emperor of the time.

Bubonic plague would then seem to have disappeared from our planet for eight whole centuries, before reappearing with devastating consequences in the Black Death of 1347–50. Thereafter, the disease smouldered with minor outbreaks until the mid-seventeenth century when for two centuries it seemed once again to have died out, only for it to reappear in China in 1894. In India, it killed some 13 million people in the years up to the First World War.

One modern bacterial disease that is difficult to explain except by vertical incidence is whooping cough or *Pertussis*. *Pertussis* has for long been known to occur in cycles of about 3.4–3.5 years, which used to be explained on the density of susceptibles theory, namely that after children susceptible to the disease become exhausted by a particular epidemic it was then supposed to take about three and a half years for new births to rebuild the density of susceptibles to the level at which a further epidemic would run. Thus, the periodicity on this theory should have been a function of population density, with the shorter periods being found in inner city areas of very high density, but periodicity was found to be everywhere the same, in town and country alike, and from country to another. If the standard theory had been correct, the sudden reduction in the density of susceptibles brought about in the 1950s by the introduction of an effective vaccine should have greatly disturbed the 3.4–3.5 year cycle, or even destroyed it altogether. Yet the periodicity persisted exactly as before, but with the total number of cases much reduced.

Perhaps the strongest evidence of all for ongoing incidence from space relates to the case of influenza. At the end of the nineteenth century, before the viral nature of influenza was established, the distinguished English physician Charles Creighton asserted that the spread of the disease could not be explained on the basis of person-to-person infection. The data from several epidemics had shown him that outbreaks occurred almost simultaneously over great distances, although the spread over localised regions took place in a more leisurely manner.

Perhaps the most disastrous influenza epidemic in recent history occurred in 1918–19 and caused some 30 million deaths. After carefully reviewing all the available information about the spread of influenza during this epidemic Dr Louis Weinstein commented thus:

> Although person-to-person spread occurred in local areas, the disease appeared on the same day in widely separated parts of the world on the one hand, but on the other, took days to weeks to spread relatively short distances. It was detected in Boston and Bombay on the same day, but took three weeks before it reached New York City, despite the fact that there was considerable travel between the two cities. It was present for the first time at Joliet in the State of Illinois *four weeks* after it was first detected in Chicago, the distance between those areas being only 38 miles.

During the same pandemic in January 1919 Governor Riggs of Alaska reported to a committee of the US Senate that influenza had spread all over an area the size of Europe and with only a small thinly spread population of about 50,000. This was despite conditions for human travel being worse than anybody could remember:

> The territory has to be reached by dog team. You have the short days, the hard, cold weather, and you only make 20 to 30 miles a day. The conditions are such as have never happened before in the history of the territory . . .

Here we have a clear case where influenza was not spread by person-to-person contact. The theory that birds harbouring the new virus may have spread the disease by spraying bird muck into the atmosphere is also disproved because flocks of birds did not fly over Alaska during the freezing winter in November and December. But the winter jet stream could well have overturned the upper atmospheric air, causing a cloud of virus-laden particles to descend on even so large an area as Alaska. Then again the disease could not have been spread from Boston to Bombay in a day. Nor could the fastest flying birds make that journey in the time. Nor are there winds that blow over such a route, with such a speed and over such a distance. But a virus or a trigger for it embedded in micrometre-sized particles, falling through the high atmosphere, will in general arrive at ground level at different places at different times. There will always be a place where it arrived first, and this could be where a new epidemic is seen to start. That a spaceborne pathogen could touch down simultaneously at two such widely separated places as Boston and Bombay does not strain credulity at all. On the other hand, it is exceedingly difficult to believe that anything travelled horizontally between these two afflicted areas.

Thirty years on, and it was the same story all over again. The worldwide epidemic of 1948 apparently first appeared in Sardinia. A Sardinian doctor, Professor F. Margrassi, commenting on this writes:

We were able to verify the appearance of influenza in shepherds who were living for a long time alone, in solitary open country far from any inhabited centre; this occurred absolutely contemporaneously with the appearance of influenza in the nearest inhabited centres.

One of the most striking features throughout this whole story is that the technology of human travel has had no effect whatsoever on the way that influenza spreads. If influenza is indeed spread by

145

contact between people, one would expect the advent of air travel to have heralded great changes in the way that the disease spreads across the world. Yet the spread of influenza in 1918, before air travel, was no slower and no different from its spread in more recent times.

It is commonly held that person-to-person transmission of influenza is proved by the very high attack rates in institutions such as army barracks and boarding schools. The efforts of Fred Hoyle and myself to test this proposition has now stretched over two decades. In the early months of 1978 we conducted a survey of boarding schools in England and Wales during an epidemic of the so-called Red Flu – a virus type (H1N1) that had been absent in the human population for some 20 years – and was suddenly raging at this time. Children under the age of 18 could surely not have encountered this virus during their lifetime and were therefore equally susceptible to the disease. We saw these conditions to be ideal for testing the hypothesis that the virus or some biochemical trigger for it was falling through the skies. Our sample involved more than 20,000 pupils with a total number of some 8,800 victims. The distribution of attack rates showed wide variations from the average 44 per cent attack rate. Very many schools showed much lower attack rates and only three schools out of more than 100 had attack rates in excess of 80 per cent that have been claimed to be the norm.

If the virus responsible for the 8,800 cases were passed from pupil to pupil, much more uniformity of behaviour would have been expected. From the spread of the attack rates across the country we saw evidence for great diversity, with a hint that the attack rate experienced by a particular school (or a house within a school) depended on where it was located in relation to a general infall pattern of the virus or its trigger on to the ground. The details of this infall pattern would of course be determined by local meteorological factors. The infall clearly displayed patchiness over

146

a scale of tens of kilometres, the typical separation between the schools.

One of the schools in our survey, Eton College, had conditions that were particularly well suited for testing the doctrine of person-to-person transmission. There were 1,248 pupils distributed in a number of boarding houses and the total number of cases across the whole school was 441. There was a huge variation in the attack rate house by house. One particular college house with a total population of 70 had only one case, compared with the expected value of 25 on the assumption of random distribution, in a person-to-person infected model. Another house with a total population of 50 had 32 cases compared with an expected value of 18 on the basis of an infective model. Here again we see patchiness, but now on the scale of hundreds of metres. This entire distribution would be expected once in 10^{16} trials on the basis of person-to-person transmission. Clearly, if one looks objectively at the facts influenza cannot be seen to be transmissible from person to person as our present-day scientific culture would have us to believe.

Further evidence against the standard dogmas of influenza transmission comes from a study in Japan. Japan is remarkable in that several large areas have population densities in excess of 2,000 per square kilometre, whereas others have well under 200 per square kilometre. Standards of monitoring and reporting disease are also uniformly good throughout most of Japan. Through the good offices of ShinYabushita we were able to obtain annual influenza notification rates in each of 47 prefectures (administrative units) into which the country is divided. We had also estimates of the total population densities in each of these prefectures. Notification rates are only a small fraction of course of the total case rates, but on average a proportionality between these indices are to be expected. Table 1 shows the relevant data for these prefectures over five years. The incidence rates are exceedingly patchy from one prefecture to the next in any given year, demonstrating an 'Eton

Table 10.1 Notification of influenza by prefecture, per 100,000
(The year by year columns give for each prefecture the number of
notifications × 100,000 ÷ population in the prefecture)

Prefecture No.	Population per square km	1970	1977	1979	1980	1981
8 (Ibaraki)	423	6.6	61.4	1.0	16.8	2.5
11 (Saitama)	1580	114.4	1.4	–	0.8	0.1
12 (Chiba)	869	37.4	5.5	–	6.4	4.2
13 (Tokyo)	3600	63.4	2.8	0.4	1.1	0.6
14 (Kanagawa)	2197	51.7	20.5	6.7	5.7	1.4
15 (Niigata)	205	113.3	2.0	5.4	25.2	6.3
16 (Toyama)	225	67.5	312.1	2.5	101.6	4.9
17 (Ishikawa)	217	339.7	95.9	139.1	10.8	0.4
18 (Fukuoki)	117	375.5	2850.4	–	57.1	264.4
19 (Yamanashi)	189	126	19.1	1.0	16.5	0.4
20 (Nagano)	165	5.6	49.7	5.9	3.5	–

College' effect occurring over distance scales of tens of kilometres,
the typical separation between prefectures.

Both high population density prefectures and low population
density prefectures show exactly the same result, an effect that is
markedly in conflict with the person-to-person transmission model.
Here is yet another fact that supports the idea that the virus, or at
least a trigger of it, has an incidence that is controlled by meteorological factors. It falls from space and its descent from the stratosphere to ground level depends on global circulation patterns of the
atmosphere.

Such a process also accounts for the otherwise inexplicable
phenomenon that epidemics of 'flu tend in general to occur with a
distinct seasonality in widely separated parts of the world. In
normal years the northern hemisphere 'flu season in temperate
latitudes tends to peak in the winter months from December to
February, whereas in the southern hemisphere, for example in

Australia the peak is 6 months displaced in June–August. In the tropics on the other hand, for example in Sri Lanka, influenza can strike at any time of the year and there is no seasonal effect. When new strains of the influenza virus arise it is mind-boggling, in terms of the normal theory, to understand how the disease could be contained within the confines of seasonal winter peaks in the two hemispheres, despite the heavy traffic of airline passengers, for instance between Australia and Europe. If a new strain of the virus emerges first in the East, surely one would expect it to shoot across into Europe with the first plane carrying an infected passenger or passengers.

According to the point of view developed in this chapter, reservoirs of the causative agent for influenza are periodically re-supplied from a comet or comets at the very top of the Earth's atmosphere. Small particles of viral sizes or smaller tend to remain suspended high up in this region for long periods unless they are pulled down into the lower atmosphere. In high latitude countries, such breakthrough processes, where the upper and lower air become mixed, are seasonal and occur mostly during the winter months. Thus a typical influenza season in a European country would occur between December and March as we have seen. Frontal conditions with high wind, snow and rain effectively pull down viral-sized pathogens close to ground level. The complex turbulence patterns of the lower air ultimately control the fine details of the attack at ground level, and determine why people at one place and at one time succumb, and why those in other places and at other times do not.

There was a direct demonstration that the general winter down-draft in the stratosphere occurs strongly over the latitude range 40 to 60 degrees in the last of the series of atmospheric nuclear tests carried out in the middle of the last century. A radioactive tracer, Rh-102, was introduced into the atmosphere at a height above 100 kilometres and the incidence of the tracer was then monitored year by year through aeroplane and balloon flights. The radioactive tracer took about a decade to clear itself through repeated down-drafts that occurred with a distinct seasonality. The downdraft was found to be much greater in temperate latitudes than elsewhere, with the period January–March being the dominant months. These observations agree with the well known winter season of the viruses responsible for the majority of upper respiratory infections, including influenza.

We have dwelt at length on influenza for the reason that here we might have the most direct evidence of ongoing incidence of biological material from space. There is an unending list of other diseases that are also likely to be driven from space. The year by year incidence of RS (respiratory syncytial virus) bears further evidence of a seasonal winter effect. The peaks of RS incidence coincides with the peaks of influenza, as indeed does a host of unnamed and unidentified viruses that come to the attention of General Practitioners with every winter season.

There has been much recent interest in the appearance of clusters of disease caused by relatively common bacteria that suddenly and mysteriously acquire enhanced virulence. Such clusters could arise when clumps of pathogen riding within small cometary objects (perhaps of the order of tens of centimetres in size) get through the atmosphere and are dispersed near ground level. There can be little doubt that this could happen from time to time. A case in point could be meningococcal meningitis. This disease has a sporadic low level of activity throughout the year in the UK, but significant increases of incidence are noted during the winter months. A mystery associated with this disease is the appearance of geo-graphically localised clusters that can persist for weeks, months or even years. Although the causative bacterium *Neisseria menin-gitidesis* that is isolated from patients is indigenous to the human population and may not be currently incident from space, an agent that transforms this bacterium into a potentially lethal variant could indeed have a space incidence that is geographically localised. The agent could for instance be a plasmid, a virus affecting the bacterium.

The conventional wisdom is that the meningococcal meningitis is propagated only through intimate personal contact, by means of droplet infection. Yet in all the best documented instances of clusters of the disease, unequivocal evidence to support this claim is sadly lacking. In an outbreak that occurred in a Cardiff University hall of residence during November and December 1996, the confirmed cases (including two deaths) were apparently not in social contact with each other. Nor were the Cardiff cases linked in any way to other cases reported up and down the United Kingdom at almost the same time. There was no evidence of direct person-to-person spread, and the incidence of the disease seems more to be connected with the location where individual victims lived or worked, rather than the people with whom they were in contact. Again, for a disease with a very low overall incidence rate, the situation was no different from the case of influenza that we have already discussed.

Variants of the extremely common meningococcal bacterium *Neisseria meningitidesis* Group B or C appear in every instance of serious disease. The indications are that an airborne, vertically incident plasmid could be involved in transforming a normally benign *Neisseria meningitidesis* into a potentially lethal variant. If this is so the 'culprit' of meningococcal outbreaks could be seen as the transforming agent or plasmid rather than the bacillus.

A generally similar story might relate to the outbreaks of *E. coli* poisoning that continue to make the headlines from time to time. The bacterium *E. coli*, like *Neisseria meningitidesis*, is normally a harmless microorganism. An airborne plasmid could be responsible to changing this into one of its lethal forms, e.g. *E. coli 0157* such as caused havoc in Scotland in November and December 1996, and which was involved some weeks earlier in similar outbreaks in Japan, Canada and the United States.

There are also many instances of localised attacks of epidemic diseases in historical times. A great plague broke out in the ancient city of Athens in the year 430BC. The Peloponnesian War that marked the decline of Athens commenced a year before. The chronicle of this war was written by Thucydides with what has been described as minute and scientific accuracy. He describes the plague in such great detail that many a modern physician has been tempted to try to identify the disease from clinical symptoms. This has proved exceedingly difficult. Some have tried to link the disease to smallpox, but in this case it should have spread like wildfire and not have ended as abruptly as it did. After reviewing various alternative diagnoses one medical commentator of the last century wrote thus:

I have looked into many professional accounts of this famous plague, and writers, almost without exception, praise Thucy-dides' accuracy and precision, and yet differ most strongly in the conclusions they draw from the words. Physicians – English, French, German – after examining the symptoms, have decided it was each of the following: typhus, scarlet, putrid, yellow, camp, hospital, jail fever; scarlatina maligna; the Black Death; erysipelas; smallpox; the oriental plague; some wholly extinct form of disease . . .

Amid this confusion it would be safe to conclude that it was unlike any disease that was known before or after this time. It was localised around Athens, appears suddenly from nowhere, an dis-appears equally suddenly and mysteriously.

A medical puzzle in more recent historical times involves a group of Trio Amerindians who for a long time had lived in isolation from the rest of humanity. When the forests in South America were cleared in the early part of the last century a tribe of 500 Trio Amerindians were discovered by anthropologists. It was found that there were several victims of polio amongst them and it transpired that they had contracted the disease at times coincident with epidemics in cities hundreds of miles away. There was no conceivable way by which forest-dwelling Surinam Indians could have contracted polio from city-dwellers. But the city-dwellers and the Indians could both have caught the disease if the causative virus (or a trigger for it) rained on them from above.

The attack of pathogens from space are of course not confined to man. There is a considerable body of evidence relating to attacks on non-human species as well. For example, a new viral disease began to take a heavy toll on dogs in the year 1978. The most remarkable feature here was that this lethal disease appeared almost simultaneously in widely separated parts of the world. Veterinary surgeons in 22 states of America, in Canada, South Africa, Holland and Australia all noticed the disease to appear for the first time at about the beginning of May of that year. Since there is clearly no traffic in dogs between these countries, and because the most stringent quarantine measures were everywhere quite ruthlessly enforced, this simultaneity of attack is strongly suggestive of an invasion from space.

In another remarkable story Grey seal in Lake Baikal in Central Siberia were found some years ago to be dying in considerable numbers from a respiratory virus. Some six months later seals in the North Sea were found to be dying of the same virus. How was this possible by victim-to-victim transference? The ridiculous answer offered by conventional wisdom is that a seal must somehow have managed to swim from Lake Baikal to the North Sea. But Lake Baikal has no outlets to the sea.

Foot-and-mouth disease is an affliction of ruminant animals that has presented itself with a baffling set of facts at the time of writing. Attempts to explain its emergence after 34 years in Britain and other parts of Europe have so far met with little success. This is of course a disease that is endemic in many parts of Africa and

Asia, and for it suddenly to cause havoc in Britain in 2001 must call for an exceptional set of conditions. The causative agent is reckoned to be so exceedingly infective as to justify closure of footpaths, cancellation of sporting fixtures and even a postponement of a General Election. If so, one has to explain the strict confinement of the virus for three decades or more to localized regions of the planet where the disease remained endemic, despite the huge volume of traffic of people and commodities that must surely have crossed these regions and into Europe on a regular basis. Despite the draconian measures that were taken to cull large numbers of healthy animals to act as a fire wall around infected farms, the disease continued to spread, often appearing in isolated parts of the British Isles. It is worth noting that this outbreak began during a spell of bad winter weather, as did the outbreak at the end of 1967. As with many human respiratory diseases a winter effect such as is observed here is suggestive of a fall-out of some viral trigger through the atmosphere. The cosmic dragons may have got the cows and sheep in our beloved countryside after all.

Those who feel uncomfortable with the logical inferences drawn from a theory that seemingly reinstates an ancient superstition often seek eye-catching one-line 'disproofs'. One objection that is often raised is that viruses and viral genes cannot conceivably be replicated independent of host cells within comets. That may be the case for a fully-fledged virus particle exuded from an infected plant, animal or human. But it is not true if viral genes were included (possibly in pieces) in the genomes of autotrophic microbial cells that *would* be able to replicate within comets. The viral genes would then multiply as the microbial cells divide, leading to a huge cascade of copies. The influenza virus, as indeed other pathogenic viruses, would then arrive in pieces that penetrate prospective host cells, and the assembly of fully-fledged viral particles would occur within the cells of terrestrial life-forms. Another objection is that viruses, whether they come in pieces or not, are host-specific, and so it is claimed must have evolved in close proximity to the terrestrial species they attack. In other words, the critic asks: 'How could the incoming virus or its components know ahead of its arrival the nature of the hosts that may be present?' The answer is simple and one would hope decisive: the virus could not of course anticipate us, but we the host species could anticipate the virus, since we must have had a long and continuing exposure to the viral genetic sequences of a similar kind, all of which had an origin in comets. Viruses are known on occasion to add on to our genes, and so viral DNA sequences would serve as invaluable evolutionary potential. Our genomes would then be made up of cometary viruses, a situation that is not inconsistent with available biological data. If the influenza virus or one that is similar to it forms part of our genetic heritage, then the so-called host specificity, or the apparent human-virus connection is instantly and elegantly explained.

At first sight it might seem as if it would have been advantageous for terrestrial organisms to have evolved towards complete immunity towards all forms of external viral invasion. The fact that evolution over 4,000 million years has not proceeded in this way must, I believe, have a profound significance. It is not hard to see how the properties of an existing gene might be changed to a minimal degree, or how an existing gene might be divided into two or more genes, processes that are of course known to occur. But the development of an entirely new gene is not so easily explained.

On this picture we could easily understand why evolutionary progress occurs by leaps, corresponding to the addition of new genes, rather than by the smaller changes that are usually considered. It now becomes clear that terrestrial organisms could not have cut themselves away from the invasions of cometary viruses. Had this happened evolution would have come to a grinding halt. Cometary dragons may breathe death on occasion, but they also help us to progress towards greater and better things in the long term.

155

References

ANDREWES, CHRISTOPHER (1965). *The Common Cold*, London: Weidenfeld & Nicholson.

HOYLE, F. & WICKRAMASINGHE, N.C. (1979). *Diseases from Space*, London: J.M. Dent.

HOYLE, F. & WICKRAMASINGHE, N.C. (1980). *Evolution from Space*, London: J.M. Dent.

KOLATA, GINA (1999) *Flu: The story of the great influenza epidemic of 1918 and the search for the virus that caused it.* London: Macmillan.

11. Dragons Bring in the Shivers

Winter desolation:
In a world of one colour
The sound of the wind.
Matsuo Basho *(1644–94)*

In most regions of the Earth any snow that falls during the winter season thaws in the following spring and summer. This is not the case, however, in extreme northern and southern latitudes and on very high mountain tops in any latitudes. Here snow accumulates year after year, leading to the formation of a permanent snow field. In Antarctica in the southern hemisphere and in Greenland in the northern hemisphere, permanent glaciers have persisted since the end of the last ice age 10,000 years ago. Such ice-covered terrain accounts for about 10 per cent of the entire land surface of the Earth.

Twenty thousand years ago, in the midst of the last ice age, the global ice cover was higher than it is now, nearer 25 per cent. In the northern hemisphere ice sheets extended through much of North America and Europe. Most of the British Isles was covered in several hundred metres of ice, and a similar situation prevailed in North America almost as far south as New York. Evidence of glaciation also extends to the equator, on tropical mountains such as Mauna Kea and Mauna Loa in Hawaii and Mount Elgon in Uganda,

157

mountains that are now totally free of ice. Antarctic glaciation was also considerably more extensive 20,000 years ago than it is today.

One of the major advances in Earth sciences of the twentieth century was the discovery that the Earth has switched repeatedly between cold and warm climatic conditions. The most detailed and best documented studies of the Earth's average temperature over the past 2 million years reveals fluctuations by as much as 10°C that occur cyclically over timescales of about 100,000 years. A period of glaciation lasts on the average for about 100,000 years and this is generally followed by a much briefer warm period – an interglacial – that lasts typically for about 10,000 years.

Ice age conditions are known to have been dry and dusty. The dry atmospheric condition implies a low rate of snowfall, but as long as snow does not melt in the summer, this does not prevent the growth of large glaciers. During an ice age snow accumulates, albeit slowly, and snow fields become progressively thicker. Eventually the fallen snow becomes so compressed and compact that it turns into solid ice. Ever-thickening glaciers accumulating on the slopes of jagged mountains eventually overcome the friction that held them stationary and begin to slide downhill. The slowly sliding glaciers gather fragments of rock on the way, fragments ranging in size from boulders to fine dust. Like gigantic blocks of sandpaper these sliding glaciers smooth out the jagged protrusions of softer rock on the slopes of mountains and eventually gouge out deep valleys below. Through successive ice ages a mountainous terrain is glacially sculptured this way. The end result of such a process is to be seen for example in the graceful contours of the English Lake District. The hardest granite rocks of over 400 million years ago that form many of the principal peaks have been mostly untouched by glacial processes, but the softer rocks of the lower hills have been moulded through glacial action resulting in a spectacular landscape of gently curving mountains, deep valleys, lakes and waterfalls. In contrast, my native country of Sri Lanka lies close to the equator and has not been engulfed in ice-age conditions, for at least the past 600 million years.

In earlier chapters we saw how the end of an ice age could be mediated through a single cometary impact. We also saw how short-lived climatic downturns, such as happened around AD540 could have been due to episodes of cometary dusting of the Earth's upper atmosphere. This dusting could have arisen either from the Earth ploughing through a small dense patch of cometary dust or by small cometary bolides actually crashing on our planet and thus lifting up a of dust into the stratosphere. A possible confusion could arise here with other purely terrestrial sources of dust in the stratosphere. Major volcanic eruptions such as Krakatoa in 1888, are known to produce short-lived climatic disturbances. But for events in the distant past there are other discriminatory tests one could use. A period of volcanic dusting would be associated with acid rain that would leave its mark as an acid signal in the ice that formed in the glaciers of Greenland. For the AD540 event the claim has been that no such acid signal is discernible in ice drills that go back to this time, and so the impact hypothesis seems more likely. Micrometre-sized ice or organic dust particles in the stratosphere can affect the climate by providing a reflective blanket around the Earth. The blanket reflects part of the Sun's radiation back into space, less sunlight reaches the surface, and the Earth cools. There is of course a reflective blanket of such particles at all times, and this is strikingly evident in the long duration of summer twilight in high latitude regions of the planet. The phenomenon of twilight is particularly affected by the scattering of sunlight by particles of submicron size. Above an altitude of 50 kilometres a whole megatonne of icy submicrometre-sized particles can be shown to exist, but this is normally not sufficient to have any effect on climate balance. A 100-fold increase in this dust loading, that could occur easily during a cometary dusting episode could, however, lead to a cooling of the Earth by 3–5 degrees C, enough to cause a mini-ice age or a climatic downturn. Such a cooling does not immediately plunge the Earth into a full-blown ice-age because of the heat currently stored in the oceans that is equivalent to the supply of sunlight over an interval of three to five years. This oceanic heat store offers a buffer for any short-term deficit of energy input from the sun.

The much spoken-of and greatly maligned Greenhouse Effect was proposed as early as 1864 by John Tyndall who, we saw earlier, was also an early proponent of the theory of panspermia. The greenhouse effect is of course a vitally important factor in controlling climate. Without it the Earth would be about 40 degrees colder than it is now, and we would be locked in a perpetual ice age. The main gases in the atmosphere responsible for maintaining the greenhouse effect are water vapour and carbon dioxide, and to a lesser extent methane. Of these it is water vapour that turns out to be the most important component at normal times. Water vapour in the atmosphere is derived from the evaporation of the oceans that then comes down as rain. The evaporation rate depends on the temperature. The present-day annual evaporation and precipitation rates of water are reckoned to be about 80 centimetres on average, sufficient to maintain the greenhouse effect as it cycles through the atmosphere and stratosphere, and back into the oceans.

Suppose now that the Earth were suddenly to be shrouded by a blanket of reflective dust that was maintained throughout a long enough period to plunge the planet into an ice age. The greenhouse effect would be diminished as soon as the backup heat store in the oceans was exhausted. The oceans would then cool and the evaporation of water would decline. When the average surface temperature falls by 10 degrees C, glaciers would grow at a steady rate, the ice cover of the planet would increase from 10 per cent to 30 per cent and we would have a full-blown ice age. The entire surface of the Earth would reflect more sunlight and ice-age conditions would continue in a stable way even after an initial blanket of reflective particles disappeared. In such a condition the Earth would

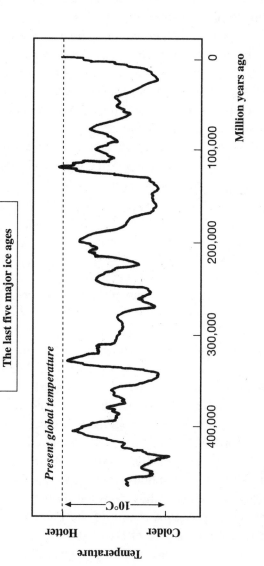

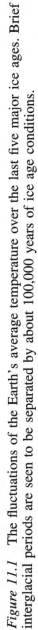

Figure 11.1 The fluctuations of the Earth's average temperature over the last five major ice ages. Brief interglacial periods are seen to be separated by about 100,000 years of ice age conditions.

remain locked in an ice-age state for an eternity, unless some dramatic perturbation were subsequently to occur.

Let us see what actually happened throughout the past history of the Earth. The geological record is divided into several ice epochs, and each ice epoch extends over a period ranging from 25 to 200 million years. Over the past 1 billion years there have been four well-defined ice epochs, but the best documented evidence relates only to the most recent ice epoch during the so-called Cenozoic era. Figure 11.1 shows the temperature fluctuations that occurred during the last few cycles of the most recent ice epoch. The emergence from the last ice age into the present interglacial state was also characterised by several sharp oscillations of temperature, but over much shorter time intervals. In detailed studies of ice precipitation rates, such as those conducted by W. Dansgaard, J.W.C. White and S.J. Johnsen, it is found that a first tentative rise of temperature took place about 15,000 years ago. But this warm spell did not last and the Earth quickly drifted back into glacial conditions. The temperature rose again at about 11,500 years and thereafter oscillated a few times before a decisive emergence from the ice age took place nearly 10,000 years ago. There is evidence that the final transition took place with great speed, possibly over decades. We attributed this final emergence in an earlier chapter to the impact by a kilometre-sized comet. The impact would bring in enough water to restore the greenhouse effect. It could also trigger eruptions of volcanoes and earthquakes of the kind described by Edward Gibbons as events that happened on a lesser scale in the middle of the sixth century AD. Such eruptions would add carbon dioxide and methane to the atmosphere, thus boosting further the restored greenhouse effect. A sudden warming of the Earth following the impact could also have had the effect of unleashing a vast store (100 megatonnes) of methane that lay trapped in permafrost, continental margins and on the ocean floor. Some of the methane would have burnt near the ground to produce more carbon dioxide. Methane rising into the stratosphere could have been oxidised to

162

form water, and water crystals forming at great heights would have reflected sunlight, cooled the Earth if it got too hot, thus acting as a thermostat and so preventing a runaway greenhouse effect.

The earliest astronomical theory of ice ages on record dates back to the work of James Croll in the 1860s. Croll's work went unnoticed until it was later revived and revised by Milutin Milankovitch in the 1940s. The Croll–Milankovitch theory of ice ages depends mostly on the fact that the axis of spin of the Earth wobbles in the manner of a child's top. The periodically changing tilt of the Earth's axis results in changes of the relative distribution of sunlight over the two hemispheres, but the total amount of energy intercepted by the planet cannot change. The wobbling axis could at best produce cooling only in one hemisphere, and yet we know that ice ages occur contemporaneously in both hemispheres. Other astronomical theories that involved changes in the Sun's energy output are also flawed. Whilst the luminosity of the Sun can change slightly over long periods of time, it cannot oscillate over short timescales as the ice-age data would imply.

Given the extreme stability of the ice-age condition, it is difficult to imagine how an ice age, once it is established, can ever be disturbed by a process that' is internal to the Earth. A kilometre-sized comet impact of the type we considered earlier could have snapped the planet out of its frozen state 10,000 years ago. The curve in Figure 11.1 would require similar events to have taken place over the past million years, separated one from another by an average interval of about 100,000 years. This is not an unreasonable estimate for the average interval between collisions by comet fragments of sizes in the range 0.5–1 kilometre.

From past record it would seem inevitable, that the cosy inter-glacial in which we now live would some day come to an end, and if 10,000 years is taken as the average duration of an interglacial period, we may be living on borrowed time. Climate records over the last 10,000 years show a random drift of temperature, upwards and downwards, by about a degree C per century. It is therefore

163

clear that a situation would eventually arise when the number of downward steps accumulates to 10 degrees C, at which point a full-blown ice age will be inevitable. The Earth would fall back into the ice-age trap, and only a further cometary bolide impact would snap it back into an interglacial.

While the glaciations that occurred in the present ice epoch involved the icing-up of no more than a quarter of the Earth's surface, more extreme episodes of near total glaciations may well have taken place in earlier epochs. There is a growing body of evidence to point to the occurrence of an extreme episode of glaciation about 700 million years ago. The first hint of such a 'Snowball Earth' during the late pre-Cambrian came with the work of Brian Harland in the 1960s. Harland studied the worldwide distribution of 'dropstones' – petrified assemblies of glacial boulders and silt, rocks that are carried along with drifting glaciers and which sink to the seabed when the glaciers melt. Wherever mountain ranges have risen from such seabeds, 'dropstones' can be discovered at the surface, and the precise dates of the glaciation in which dropstones formed can also be estimated from isotope studies. Dropstones are found all over the world, but what baffled Harland and others was the presence of stones that could be dated at about 700 million years at locations near the equator. A particularly rich harvest of tropical dropstones was recently found in Namibia in south-west Africa.

Could one avoid the conclusion that glaciation could have extended all the way down to the equator 700–600 million years ago? One possible way out was to invoke the operation of Wegener's theory of Continental Drift. There are arguments for supposing that all the present-day continents were indeed clustered around the south pole 700 million years ago. One could then go on to assert that Namibia was nowhere near the equator 700 million years ago, and so a global glaciation event might be avoided. This possibility has evidently been ruled out by Joe Kirshvink on the basis of studies he has made of the magnetism of dropstones. When the dropstones accumulated from soft magnetic silt, the resulting stones would have locked in an imprint of the magnetic field of the Earth at the location where they solidified. By studying the magnetism of the dropstones it was possible to say whether they formed near the south pole or near the Equator, and for the Namibia stones the answer was a decisive win for the tropics. If Kirshvink's arguments hold, it would seem inevitable that the big freeze of 700 million years ago extended down to the Equator, and indeed enveloped the entire planet.

Once glaciation starts to move towards the Equator the big freeze would of course be a runaway affair. When more than about 70 per cent of the surface is ice-covered, sunlight absorption is cut, the greenhouse effect is dramatically reduced, and the Earth's temperature plunges to −40 degrees C, leading to a total and irreversible freeze-up. With all the oceans frozen to depths of hundreds of metres, the life forms that existed 700 million years ago − only single-celled microorganisms, typified by cyanobacteria, that derived their energy from sunlight would have surely become extinct on a relatively short timescale, unless the ice was of an unusually transparent kind.

The main problem that faced the 'Snowball Earth' theory was that once the big freeze began it could not easily end. Kirschvink proposed a way out of this dilemma by turning to volcanoes. Volcanoes, it was argued, would break through to the surface from time to time notwithstanding the huge amount of overlying ice. Exceptionally violent volcanic eruptions occurring in rapid succession would not only produce enough energy to melt the ice, they could also unleash vast quantities of carbon dioxide into the atmosphere. It has been argued that an exceptionally strong greenhouse effect might ensue that would last for centuries, leading in turn to a vigorous meteorological cycle that caused torrential acid rain and frequent hurricanes. The effect of all this would be for the run-out of water to produce extensive deposits of calcium carbonate overlying glacial deposits. Thick carbonate layers have indeed been found lying immediately above the dropstone laden glacial sediments in Namibia, and this evidence has been sited as a vindication of the Snowball Earth theory. If that is so, there is no reason why such major global freeze-up and thaw occurred only once. According to Paul Hoffman and Daniel Schrag there is evidence to suggest that the snowball–hot-house alternation occurred not once but three times during the period between 750 million years ago and 580 million years ago. In each cycle the deep freeze state may have lasted for as long as 20 million years.

All these ideas are still being vigorously debated with every bit of evidence being scrutinised with a toothcomb. It is fair to say that the jury is still out as to whether the Snowball Earth could really have happened in the way proposed by Hoffman and Schrag. But if it did happen here are other coincidences that are worthy of note that would point strongly to the part that may have been played by the cosmic dragons. If all life became extinct at the snowball stage, how did it reappear? Reappear it surely did, and precisely at the moment when the Earth emerged from the last oscillation in the deep-freeze–hot-house cycle 570 million years ago.

The resurgence of life at this time took a most remarkable turn. For the first time in the history of the planet we find a huge surge of multicellular life – metazoa – appearing almost in an instant. The number of phyla or body plans that appeared at this time are by all accounts greater than the number of metazoan phyla that survive today. Consistent with the theories described earlier in this book, we have here a strong indication that metazoans arrived with the flood of comets that snapped the Earth out of its deep-frozen state. It is also possible that this new crop of comets arrived in the solar system from a distant part of the galaxy, and was not part of the original complement of solar system comets that had hitherto been able to deposit only single-celled life forms on to the Earth.

The impact of a large comet with the Earth would seem an obvious way of terminating an otherwise seemingly interminable Snowball condition. Evaporation of water from the energy of the impact could have started the process of restoring the greenhouse effect that could then be taken over by a secondary triggering of volcanic and seismic activity of the kind required in the purely earthbound theory of Hoffman and Schrag.

The impact hypothesis for unlocking the deep-freeze is further supported by recent data from space exploration described in a book by John and Mary Gribbin. The Gribbins point out that the surface of Venus mapped by the space probe Magellan shows evidence of a remoulding impact event at precisely 600 million years ago. This impact would appear to have led to the melting of a large part of the Venusian surface. In another remarkable study the dating of tiny glass beads called tektites in lunar soil have shown a sudden increase in the frequency of tektite-forming impacts also at about the same time, 600 million years ago.

To sum up then, the cosmic dragons kicked our planet out of a deep-freeze state and at the same time breathed life back on to its barren shores. It is this life that eventually led to Man – a creature capable of looking back at the very processes that created it.

References

DANSGAARD, W., WHITE, J.W.C. & JOHNSEN, S.J. (1989). 'The abrupt termination of the Younger Dryas climate event', *Nature*, 339, 532.

GOULD, STEPHEN JAY (1989). *Wonderful Life*. London: Hutchinson Radius.

GRIBBIN, JOHN & MARY (1996). *Fire on Earth*. London: Simon & Schuster.

HOFFMAN, PAUL F. & SCHRAG, DANIEL P. (2000). Snowball Earth, *Scientific American*, January.

HOYLE, FRED (1981). *Ice*. London: Hutchinson.

HOYLE, FRED & WICKRAMASINGHE, CHANDRA (2001). 'Cometary Impacts at Ice Ages' *Astrophys.Sp.Sci.*

12. Dragons Come Home
to Roost

*I cannot give any scientist of any age better advice
than this: the intensity of conviction that a hypothesis
is true has no bearing on whether it is true or not.*
P.B. Medawar, *Advice to a Young Scientist, 1979*

Whilst many may still despair of the sanity of the idea of plagues
and evolution being driven from space, the consensus view of the
scientific community is moving inexorably in the direction of an
extraterrestrial origin of life. Ideas that when first proposed were
thought to be wildly heretical are quietly slipping into the realms of
orthodox science. A major paradigm shift comparable in importance to the Copernican revolution of the sixteenth century seems
well underway.

In the mid 1970s we considered interstellar clouds to be the
chemical factories for producing the organic building blocks of life,
polysaccharides (chains of sugar molecules), amino acids (components of proteins) and even protocellular structures. Nowadays
the injection of chemical building blocks from comets to the Earth
is taken for granted, and in a very recent publication a group of
NASA scientists have argued that organic molecules found in
comets may have been derived straight from the molecules found in
the dust clouds of deep space: a complete turn around to 1977 and
before. In 1977 Harry Abadi and I wrote a letter to *Nature* (vol.

267, 687, 1977) arguing for the presence of biological molecules in Martian dust clouds, and in 1986 Richard Hoover and a group of us at Cardiff University made out a case for microbial life in the subsurface oceans of Europa. All these ancient dragons are at long last coming home to roost.

One of the great advances of biology in the past decade has been the development of techniques to map the sequences of bases in RNA or DNA. Using such maps it is in principle possible to construct phylogenetic trees, assuming that evolution from one bacterial type to another really did occur on the Earth. Trees constructed thus far have been found to be fraught with logical inconsistencies as we have hinted in an earlier chapter. Furthermore, bacteria discovered in the fossil record, judged from their morphologies, show no evolutionary differences from their modern counterparts. The oldest fossilised bacteria of 3.5 billion years ago show no changes from microbes of recent times. Bacterial evolution on Earth thus seems to be seriously in doubt except in so far as very minor changes corresponding to single point mutations and gene transfers are concerned. A bacterium, called *Aquifex aeolicus* that Iives in hot springs at temperatures close to boiling was thought until recently to have a decisively greater antiquity than other terrestrial archaebacteria. But this conclusion has come to be questioned after a genetic map of this bacterium became available. *Aquifex* contains only one gene that is not found in normal bacteria, implying that a switch between heat-loving and normal types might be a more trivial transition than was hitherto thought. From a wide range of normally occurring (incident) bacterial types, *Aquifex* just happened to be the best suited to the boiling water habitat in which it is found.

Another problem that has recently surfaced concerns the evolutionary relationship between prokaryotes (microbes without cell nuclei) and eukaryotes (microbes with cell nuclei). From gene sequence analysis it is becoming clear that the usual picture of prokayotes being more primitive cannot be entirely justified. Both eukaryotes and prokaryotes may have co-existed in the cometary mix of microorganisms from which terrestrial life emerged. In which case the presence of cometary viruses that replicate within eukaryotes could also be justified.

The public have been attuned to the idea of life on Mars ever since Giovanni Schiaparelli (1835–1910) had rashly claimed the existence of canals on this planet. Mars was therefore one of the first objectives of space exploration following the dawn of the Space Age, and the priorities of several international space agencies still continue to be focused on this planet.

A search for primitive life forms on Mars was carried out by the US space agency NASA in the year 1976. Two spacecraft named Viking 1 and Viking 2 arrived at chosen spots on the Martian surface on 20 July and 3 September 1976 equipped to make *in situ* tests for bacterial life. The presumption was that any micro-organism which may have been present had metabolic processes broadly similar to those of Earthly microorganisms. In one experi-ment a nutrient broth of the sort that is normally used to culture terrestrial bacteria was poured on to a sample of Martian soil, and, alas, gases were found to froth out as would be consistent with the presence of microbial life. However, another experiment that sought to look for organic residues left by bacteria produced a negative or doubtful outcome. The two sets of experiments seemed to produce inconsistent results.

In 1976 NASA made a categorical announcement that the Viking experiments did not support the presence of bacterial life on Mars. Such a result accorded with the reigning paradigms in science, and so the matter may well have rested for all time, but it did not. Gilbert Levin, the principal investigator of this project, has always remained uneasy about the conclusion that appeared to be so quickly reached. In 1986 a careful re-examination of the Viking data was carried out by a team led by Gilbert Levin. Their conclusion was that Viking experiments of 1976 may well have found evidence for microbial life on Mars. Such life could still be lurking in subsurface niches on the planet. Not only was it the case that the Viking experiments had not been tested beforehand on the most inhospitable terrain on the Earth – the dry valleys of the Antarctic – until well after 1976, but when they were so tested results identical to the Martian results were obtained. The Antarctic Dry Valleys surely contain populations of microorganisms, but their turnover rate was so small that no organics were detectable by the Viking instruments. Gilbert Levin and Patricia Straat further claimed that all attempts to mimic the Viking results using non-biological models were unsuccessful. Even more controversially Gilbert Levin argued that a series of colour photographs of a Martian terrain near the Viking lander site, taken at regular intervals through a Martian year, showed seasonal changes in the colours of rocks. Such results might have been consistent with the seasonal growth of microbial ecologies, similar perhaps to lichens. Be that as it may, what was shown beyond a shadow of doubt was that the results of the 1976 Viking experiments were *not inconsistent* with the presence of microbial life on Mars. These statements were of course not readily conceded at the time. Gilbert Levin has since left NASA but has consistently maintained this position, operating as an independent scientist running his own company, Biospherics Inc. Now at last, there are signs that Levin's claim to have discovered life on Mars in 1976 is being taken seriously. In future missions being planned, that include sample returns from this planet, NASA is accepting the need for level 4 biohazard containment strategies, even though the return of pathogenic microorganisms might be only a remote possibility.

Before coming to the most recent claims of microbial fossils in a Martian meteorite let's return to Earth and look at earlier claims of a similar kind, claims that were either summarily dismissed or ignored. In the mid 1960s H. Urey, and later G. Claus, B. Nagy and D.L. Europe, examined the Orguel carbonaceous meteorite which fell in France in 1864, microscopically as well as spectroscopically.

They claimed to find evidence of organic structures that were similar to fossilised microorganisms, algae in particular. The evidence included electron micrograph studies, which showed substructure within these so-called 'cells'. Some of these structures resembled cell walls, cell nuclei, flagella-like structures, as well as constriction of some elongated objects that even suggested a process of cell division. If these 'organised elements' were indeed microbial fossils, the question arises as to how such structures came to be included within carbonaceous meteorites. This question could not be satisfactorily answered at the time in 1960, although we could now say that the answer was obvious. Carbonaceous chondrites, typified by Orguel or Murchison, probably represent the residue of comets that once may have contained microbial life that was thriving within subsurface pools. These objects can thus be thought of as fragments of biological comets that have been stripped of volatiles, and within which sedimentation and compaction of microorganisms may have occurred over hundreds of perihelion transits.

One of the more serious and possibly legitimate criticisms that were made of the early microfossil claims was that the meteorite structures included some terrestrial contaminants such as rag-weed pollen. Whether this was really the case, or not, in the vast majority of examples cited by Claus, Nagy and Europe, remains unresolved at the present time. However with a powerful attack being mounted by the most influential scientists of the day, the meteorite fossil claims of the 1960s came to be quickly silenced.

In 1980 the German palaeontologist H.D. Pflug reopened the whole issue of microbial fossils in carbonaceous meteorites. Pflug used techniques that were distinctly superior to those of Claus and his colleagues and found a profusion of 'organised elements' comprised of organic matter in thin sections prepared from a sample of the Murchison meteorite which fell in Australia on 28 April 1969.

The method adopted by Pflug was to dissolve-out the bulk of the minerals present in a thin section of the meteorite, using hydrofluoric acid, and doing so in a way that permitted the insoluble carbonaceous residue to settle with its original structures intact. It was then possible to examine the residue in an electron microscope without disturbing the system from outside. The patterns that emerged were stunningly similar to certain types of terrestrial microorganisms. Scores of different morphologies turned up within the residues, many resembling known microbial species. It would seem that contamination was excluded by virtue of the techniques that were then used. No convincing nonbiological alternative to explain all the features has thus far emerged, and during the past year even more compelling evidence for meteorite microfossils has been adduced by Pflug himself, and independently by Richard Hoover, working colleagues at the NASA Marshall Space Flight Centre in Huntsville Alabama.

Of the many space projects that have captured the headlines in recent years exploration of Europa, an icy satellite of the planet Jupiter must take pride of place. Europa is slightly smaller than the Earth's Moon. In Greek mythology Europa was a Phoenician princess abducted to Crete by Zeus, the King of the Gods. As a satellite of Jupiter, Europa is subject to huge amounts of tidal flexing – enough to melt ice underneath a surface crust, leading to the formation of huge subsurface oceans. It is in the midst of such oceans that Arthur C. Clarke has placed his intelligent European sea creatures. As for the case of Mars, there is much public interest in the possibility of life on Europa. The first close pictures of Europa's surface were obtained by camera's aboard the Voyager spacecraft in 1979. Already at this time a mosaic of cracks on the surface of the icy satellite was revealed, fuelling the speculation of a relatively thin exterior crust of frozen ice overlying a warm subsurface ocean. The cracked surface is thought to arise from the effect of Jupiter's tides that rhythmically pull and push the planet.

The surface of Europa came under much greater scrutiny following the launch of the Galileo Orbiter in December 1997. Fascinating images of Europa's surface have been obtained in the past few years. Cracks, ridges and possibly even tunnels seem to show up over distance scales of thousands to hundreds of metres, and a network of cracks appears to be demarcated with organic pigments of some kind. The images of certain regions of Europa's surface resemble images of sea ice on the Earth. It seems more likely now that beneath the surface ice there is indeed an ocean as much as 50 kilometres deep, which is kept warm and liquid by tidally generated heat. If this is so it is also likely that cosmic life would have been planted here by means of cometary impacts. Scientists are currently discussing the types of future missions to Europa that would settle such matters once and for all. It is possible that answers to questions of European life might be available within a decade or two at most.

Arthur C. Clarke, perhaps the greatest futurist of our times, hazards several predictions for the next century in his book, *Greetings, Carbon-based Bipeds* (Harper Collins, 1999):

2011: . . . first robot probes drill through the ice of Europa, and an entire new biota is revealed . . .
2061: The return of Halley's Comet; first landing by humans. The sensational discovery of both dormant and active life-forms vindicates Hoyle and Wickramasinghe's century-old hypothesis . . .

The timing of 2011 for Europa sounds about right, but 2061 is in my view pessimistically too long for the absolute proof of cometary life. After all, Halley's comet is not the only comet that could be studied. Indeed one mission to a comet, the Stardust Mission, was already launched on 7 February 1999. It is due to collect material from Comet Wildt2 in the year 2004 and return this to Earth in 2006 to be examined in the laboratory for signatures of life chemicals. The experiments carried on Stardust were not primarily designed to collect living microorganisms in a way that would preserve viability. Future space missions to other comets will be aimed at collecting and bringing back material under conditions more favourable to revealing viable microorganisms. But there are always discoveries that take place unexpectedly and these are often the most interesting. As we saw in Chapter 1, the spacecraft Stardust carried an *in-situ* interstellar dust analyser (mass spectro-meter as it is called), and in April 2000 scientists analysed five interstellar dust particles that struck the surface of the detector at high speed. The particles examined were of course shattered into fragments upon impact, but the surprise was that the shattered fragments comprised of huge organic structures, structures that resembled the cell walls of bacteria.

Particles from comets could also be analysed much nearer home. We have already seen that some hundreds of tonnes of organic material of cometary origin reaches Earth on a daily basis. Clumps of interplanetary dust particles of cometary origin have been collected in the stratosphere over many years using sticky paper flown aboard U2 aircraft. These so-called Brownlee particles (named after the investigator D.E. Brownlee) have consistently shown evidence of carbonaceous material, some of which might be exceedingly complex, and some with morphologies uncannily sim-ilar to terrestrial bacteria.

In 1993, further studies by S.J. Clemett and his collaborators of eight Brownlee particles, which were identified as cometary dust, revealed the presence of exceedingly complex organic molecules including aromatic and aliphatic hydrocarbons. This discovery was clearly an early step towards identifying cometary particles as being possibly biogenic. At the present time a group of Indian scientists led by Jayant Narlikar, and including myself in the UK as co-principal investigator, is conducting experiments that involve the collection, under the most stringently sterile conditions, of large quantities of air at various heights in the stratosphere. It is hoped that the analysis of these samples in due course would answer the question whether microbes are currently coming to the Earth from outside.

The latest chapter in the exploration of panspermia was opened in August 1996 with studies of a 1.91-kilogram meteorite (ALH 84001) which is believed to have originated from Mars. ALH84001 is just one of a group of meteorites discovered in 1984 in Allan Hills, Antarctica, which is thought to have been blasted off the Martian surface due to an asteroid or comet impact some 15 million years ago. This ejecta orbited the sun until 13,000 years ago when it plunged into the Antarctic and remained buried in ice until it was discovered. The presumed Martian origin of these meteorites (also known as SNC meteorites) seems to have been confirmed by several independent criteria. One, that is perhaps amongst the most cogent, involves the extraction of gases trapped within the solid matrix which were found to resemble in their relative abundances the gases that were discovered in the Martian atmosphere. Also the ratio of oxygen isotopes $^{17}O/^{18}O$ in the mineral component matches the value found on Mars so closely that there is no reason to doubt a Martian origin.

A team of investigators led by David S. McKay have found that within the meteorite ALH 84001 there are submicron-sized carbonate globules, around which complex organic molecules are deposited. These molecules, including polyaromatic hydrocarbons, are characteristic products of the degradation of bacteria. Moreover, structures were found that are strongly suggestive of fossilised microorganisms (Plate 12.1).

McKay and his colleagues admit that their proposed identification involves a process of multi-factorial assessment. The totality of the available evidence, considered all together, points in their view to a possible microbial origin, although each single piece of evidence may be capable of more conservative interpretation. Several other groups of investigators have subsequently challenged this interpretation of the data on very similar grounds to those used in the past to dismiss claims of meteorite microfossils. 'Artifacts' or

Plate 12.1   Organic structures in the Mars meteorite ALH84001 tentatively identified as microbial fossils. (Courtesy: © NASA)

'contaminants' are the operative terms being used to denigrate any microfossil claims in extraterrestrial material.

One of the perceived difficulties with these so-called Martian microfossils was that there were far too small, about 100 nanometres across, to be identified with any known bacterial species. But thereby hangs another tale. In the past few years similar microbes with diameters of less than 100 nanometres have turned up in the most unlikely places: in deep sea sediments, in human kidney stones, and in the blood plasma of humans as well as cows. New frontiers of life, perhaps new frontiers of cosmic life, seem to be defined by a newly discovered class of microbial life. These nanometre-sized microbes are called nanobes.

Although the 1996 Martian microfossil claim is still under vigorous dispute, the concept of microbial life being moved from one planetary body to another, through impact ejection of life-bearing rocks, is rapidly coming into fashion. The Mars meteorite

ALH84001 has shown beyond any doubt that complex organic structures indigenous to a meteorite, and by inference even microbial cells, could be transferred in a viable form from one planetary body to another. Planetary panspermia, as this concept has recently come to be known, is not by any means a new theory. In his presidential address to the 1881 meeting of the British Association, Lord Kelvin drew the following remarkable picture:

> When two great masses come into collision in space, it is certain that a large part of each is melted, but it seems also quite certain that in many cases a large quantity of débris must be shot forth in all directions, much of which may have experienced no greater violence than individual pieces of rock experience in a landslip or in blasting by gunpowder. Should the time when this earth comes into collision with another body, comparable in dimensions to itself, be when it is still clothed as at present with vegetation, many great and small fragments carrying seeds of living plants and animals would undoubtedly be scattered through space. Hence, and because we all confidently believe that there are at present, and have been from time immemorial, many worlds of life besides our own, we must regard it as probable in the highest degree that there are countless seed-bearing meteoric stones moving about through space. If at the present instant no life existed upon the earth, one such stone falling upon it might, by what we blindly call natural causes, lead to its becoming covered with vegetation.

Such interplanetary transfers of life remain a possibility but one that in my view represent a relatively unimportant route for transference of life on a cosmic scale. Moreover, it fails to address the all-important question of how life got started in the solar system in the first place. According to the ideas we have discussed in this book a far better option is to have Mars, Earth and every other habitable planetary body infected with the same cometary source of life – a source of life that is derived from an even bigger system. Comets, the cosmic dragons, impact all planetary bodies, so cometary panspermia must surely remain the principal route for the transference of cosmic life.

Recent discoveries of planets around several nearby Sun-like stars which we discussed in Chapter 2 have added a further impetus to the idea that life is a cosmic phenomenon. It would of course be naive to suggest that life evolved from microbes, through multicelled life forms to higher plants and animals only on the Earth. Rather would one expect parallel evolutionary processes, starting from the same bacterial building blocks, to have occurred on every habitable planet everywhere in the Universe. Of the 100 billion or so Sun-like stars in our galaxy it would be reasonable to expect there to be, say, one per cent with planetary systems like our own, a substantial fraction of which could be inhabited by living systems. The universe would then be teeming with life.

The question of intelligent life outside the Earth continues to torment many geocentrically oriented critics. Having been forced to abandon their cherished doctrine of Earth-centred life, they now cling tenaciously to the idea that intelligence at least may be confined to the Earth. A variety of arguments are being offered in support of this position. For instance, it is argued that biological evolution proceeds only through the rigid operation of a principle of opportunism and intelligence is not envisaged as a natural outcome of such a process. The argument goes on to assert that intelligence must therefore be an exceedingly rare occurrence in the Universe. The argument fails if, as I believe, a capacity for intelligence is built into the genetic information that is widely disseminated on a cosmic scale. Intelligence on this picture would be a common occurrence in the cosmos.

If so, one may well ask where are such occurrences? Why haven't they come here or communicated with us all this while. My answer is that they are already here in our genomes. In a sense we came from space, our genomes neatly packaged within primitive cells. And all the manifest complexity of terrestrial life has a cosmic provenance. We have argued in earlier chapters that the original blueprint of this mega informational system emerged as a one-off near-miraculous event in the history of the Universe. Once it emerged its spread was inevitable and unstoppable. Whether this singular event happened through an amazing conjunction of super astronomically improbable accidents, or through the intervention of intelligent design is a question that could be asked, but cannot be resolved within the purview of empirical science. Scientists in the year 2001AD are able to perform feats of genetic engineering, albeit to a limited extent. Biotechnology companies are amassing great profits by genetically modifying crops to suit a variety of economic ends, and genetic engineering is also beginning to play a role in the field of medicine. It does not stretch credulity too far to propose that a super-intelligent entity may have actually worked out the full range of genetic arrangements that is manifest in the life forms we see around us. You could choose to call it God or Cosmic Architect if you will. This is an intellectual position that hovers uneasily between science fiction and religion, and one that is not generally held by Cartesian-minded scientists of the present generation. Whether that would change in the future is left to be seen, but this radical position does remain a logical option that cannot be denied.

According to the point of view we have developed, nothing of great innovative significance in biology has ever happened on the Earth. The Earth was simply a building site for the incomparably magnificent edifice of cosmic life. It came in units with the cosmic dragons, and occasionally went with them as well. Natural selection, according to the criterion of the survival of the fittest, selected those life forms from the cosmically available master plan that were best suited to the local environment at all times. It is against this backdrop that the normal evolutionary processes discussed in conventional biology would operate, more in the manner of fine-tuning than innovation.

On other planets around other stars these same processes would operate. Life would inevitably develop on every habitable planet, descended from those same cosmic genes, delivered by cosmic dragons. One day we shall make contact with an alien life form on a distant planet. And no matter how strange he or she or it may seem we would know that we are related to that alien creature. We have our cousins out there in the big wide cosmos. The cosmic dragons are their friends or foes as they are ours. And we would know that our genetic ancestors still lurk amidst the stars.

References

CLARKE, ARTHUR C. (1999). *Greetings, Carbon-based Bipeds*, London: HarperCollins.

CLAUS, B., NAGY, B. & EUROPA, D.L. (1963). *Ann NY Acad Sci*, 108, 580.

CLEMETT, S.J. et al. (1993). *Science*, 263, 721.

DIGREGORIO, BARRY E., LEVIN, GILBERT V. & STRAAT, PATRICIA ANN (1997). *Mars: The Living Planet*, Berkeley: Frog Ltd.

HOOVER, R.B. (1998). *Proceedings of SPIE Conference*, 3111, 115–136.

MC KAY, D.S. et al. (1966). *Science*, 273, 924.

PFLUG, H.D. (1984). In *Fundamental Studies and the Future of Science*, ed. C. Wickramasinghe. Cardiff: University College Press.

Epilogue

Throughout this book we have seen evidence of the power of the cosmic dragons. It is to them that we owe our very existence, and like it or not, life on Earth still remains precariously poised at their mercy. Put in such a way it might appear to some that we are edging dangerously close to reviving an ancient superstition: the belief that comets are endowed with mystical and ominous powers. Although entirely fortuitous, the resemblance poses a problem because it is thought that scientific knowledge must irrevocably replace ancient superstitions.

Our own ancestral links to the cosmic dragons – the comets – are, however, coming to be grudgingly acknowledged, at least to a limited extent. For increasingly, it is now thought that comets supplied the essential molecules for life's origins– the chemical nuts and bolts of life, so to speak – life itself still being considered to have originated on the Earth. I have argued throughout this book that such a limited perception is inadequate. What is required to come from the external Universe is not merely the chemical ingredients, but the entire blueprint of life, neatly packaged in the form of genes within bacteria and viruses. This cosmically derived blueprint then finds its expression on our planet, as the pieces of a cosmic jigsaw puzzle continually brought in by the cosmic dragons interlock in a myriad ways, leading to the magnificent spectrum of life-forms that we see around us.

Epilogue

Despite overwhelming evidence in support of this general thesis, resistance still lingers, and one might wonder why. One possible reason was hinted at the end of the last chapter. There is a sniff of God or Creator in the cosmic dragons and that may well be repugnant to the intelligentsia of the twenty-first century. After a long history of industrial progress and materialism that was ruthlessly driven by reductionist Cartesian philosophy, we tend to turn away from even the vaguest hint of a creative origin of life.

An intelligent Universe or a cosmic creation of life *could* be a part of the thesis of *Cosmic Dragons*, although other options remain: for instance, a Universe with no beginning or end, in which life and the cosmic dragons are ever-present. The problem of the origin of life then ceases to exist.

The option of an intelligent creation poses serious sociological problems, because the term 'creationism' has come to be used in a restricted sense to describe a religious movement in the United States that seeks to validate the biblical story of creation. From the end of the nineteenth century onwards, science and creationism have drifted apart, and modern creationist movements do indeed pose a continuing threat to science, science education and the public understanding of science. They seek to ignore the hard-won facts of science, whenever such facts conflict with the literal interpretation of a religious belief. I need hardly stress that the thesis embodied in *Cosmic Dragons* does not have a creationist or religious motive. Life ever-present in an eternal Universe, or the creation of life in a Universe with a finite timescale remain equally valid logical options, but in either case the genetic blueprint of life must be external to the Earth and cosmically derived.

The alternative to a cosmic blueprint of life arriving with the cosmic dragons is that this blueprint came to be assembled from simple chemical units. This miracle would then have had to happen against insuperable odds in a small puddle on an insignificant planet in the Cosmos. To avoid such a conclusion scientists in many laboratories are trying to cause the transformation from

183

organic chemicals to life, but needless to say that still remains a distant dream. Their confidence in this pursuit stems from the belief that if we can make machines and perform great technological feats as we do today, we must surely be clever enough to transform nonliving matter into life.

If the cosmic dragons and their well attested encounters with Earth can teach us anything, it is humility. Although we may have close to understanding where we came from, the ultimate origin of life may continue to elude us for generations to come. It may even be outside the purview of empirical science. Until the dream of our laboratory life-makers comes to be realised, and one has to admit that this may never be, the cosmic dragons must steal the show.

Index

Abadi, Harry, 168
AIDS, 141
Alfven, H, 20
ALH84001 meteorite, 176ff
Allen, D.A., 81, 83, 95
Allens Hill meteorite, 176ff
Almagast, 28
Al-Mufti, S., 79ff
Alpha Draconis, ix, 115
Alvarez, L.W. and W., 101
Anaxoragas, 37
Andrewes, Sir Christopher, 139,
    140
Aquifex aeolicus, 169
Archbishop James Ussher, 34, 113
Arfderydd, Battle of, 122
Aristarchus of Samos, 29
Aristotle, 28, 36, 88, 120
Aromatic molecules in space, 65
Arrhenius, Svante, 37, 38, 42
Arthur, King –, 121
Asaro, F., 101
Asteroid threat, 126ff
Asteroids, effect of their impact,
    130, 131
Athens, Plague of, 137, 152
Atoms, 4ff

Badway, A., 115
Bacteria in comets, 136ff

Bacteria in space, 74ff, 78, 83ff
Bacterial fluorescence, 86
Bacterial grains, extinction, 79, 80
    infrared spectra, 82, 83
Baillie, Mike, 117, 121
Basse, W. Bruce, 113
Bauval, Robert, 114
Beta Pictoris, 26
Bevan, C.W.L., 76
Big Bang Theory, 13, 14
Biological grains, 74ff, 136ff
Black Death, 143
Bless, R.C., 64
Bondi, H., 15, 74
Browne, Sir Thomas, 33, 34
Brownlee, D.E., 175
Brownlee particles, 175
Bubonic Plague, 142, 143
Burbidge, G., 12, 14, 15, 46, 75
Burbidge, M., 12, 46
Burgess Shale, 35

Cambrian explosion of life, 33, 35
Calendar of events related to
    impacts, 107
Celtic mythology, 122
Charaka Samhita, 142
Chemical history of planetary
    material, 21, 22
Chicxulub crater, 102

185

Index

Index